The
QUEST
for
IMMORTALITY

—————⋈—————

Science at the Frontiers of Aging

S. JAY OLSHANSKY

and BRUCE A. CARNES

W. W. Norton & Company   *New York · London*

For information about permission to reproduce selections from this
book, write to Permissions, W. W. Norton & Company, Inc., 500 Fifth
Avenue, New York, NY 10110

The text of this book is compose in Filosofia
Composition by Tom Ernst
Manufacturing by The Haddon Craftsmen, Inc.
Book design by Judith Stagnitto Abbate

Library of Congress Cataloging-in-Publication Data

Olshansky, Stuart Jay, 1954–
The quest for immortality : science at the frontiers of aging /
S. Jay Olshansky and Bruce A. Carnes.
p. cm.
Includes bibliographical references and index.
ISBN 0-393-04836-5
1. Longevity. 2. Aging. I. Carnes, Bruce A. II. Title.

QP85 .O474 2001

612.6'8—dc21   00-058391

W. W. Norton & Company, Inc., 500 Fifth Avenue, New York, NY 10110
www.wwnorton.com

W. W. Norton & Company Ltd. 10 Coptic Street, London, WC1A 1PU

1 2 3 4 5 6 7 8 9 0

For Sara, Jessica, and Ricky   [S.J.O.]

*For my wife Linda and our children Laurel, Rachael, and Nathan,
my brother Michael, and in loving memory of my parents
Mark B. and Donna Mae   [B.A.C.]*

Out of the darkness of the womb, into the darkness of the grave,
man passes across his narrow strip of life. Two vast
eternities stretch oceanlike on either side of the
island of individual existence, and through
the darkness that enshrouds them no
human eye, it has been thought,
could ever pierce.

———⟩⟨———

ANNIE BESANT, 1886

CONTENTS

———✕———

Foreword

Just remember, once you're over the hill,
you begin to pick up speed.

CHARLES SCHULZ

IT IS ALWAYS AN ADVENTURE when my in-laws come to visit. The day before they arrive, I am sent out in search of foods high in vitamins and fiber, and free of fat, sugar, salt, and taste. Once they have distributed gifts to their grandchildren, a large sack containing every vitamin, mineral, antioxidant and the latest "scientifically proven" antiaging substance reported on the news is emptied on the counter. Pouches of fresh oatmeal and bran are placed neatly on the kitchen counter next to containers of prunes and other dried fruits. Pointing to the countertop, my eighty-year-old father-in-law boasts that each of these dietary supplements will do nothing less than prevent heart disease and cancer. He proudly proclaims that he is feeling fit, and ready to go on his daily forty-five-minute power walk. My mother-in-law asks whether she would be better off buying copper rings or magnets to wear around her joints to alleviate the nagging pain of her arthritis. Her latest health kick is a "breakthrough" discovery that will clear out arteries and eliminate the risk of heart disease—soy protein. Variations on this scenario have been played out for over ten years.

My in-laws are alive today not because they take vitamins, eat oat bran, or ingest hormones. Both were saved by medical technologies that successfully treated the cancer and cardiovascular diseases that occurred in spite of following dietary practices they were led to believe would prevent these diseases. To their credit, they have been physically fit throughout most of their lives because they exercise almost every day.

The quest for immortality, the war with disease, and the fear of death have preoccupied humanity for thousands of years. Today is no different. Can aging-related diseases be eliminated, the aging process reversed, and an ageless body achieved, as some experts claim? Is there a longevity strategy people can follow that will lead us all down the path to a healthy and happy 100-year lifespan? Can a combination of home remedies, herbs, and antioxidants forestall aging indefinitely and serve as miracle cures for heart disease and cancer? Can you grow younger with growth hormone, experience the miracle anti-aging properties of melatonin, and find the antidote to aging by replenishing other lost hormones? Does your biological age actually decline when you floss your teeth and avoid using cell phones? Can a lifetime of caloric restriction lead to a 120-year lifespan? What does science have to say about these approaches to extending longevity? Can the science of today offer alternative strategies for increased lifespans?

The news media perpetuates confusion about aging and disease by presenting conflicting messages to the public. We are bombarded by reports that just about everything but pure spring water causes cancer—even tap water is declared dangerous by some people. In the 1960s, the late Nobel laureate Linus Pauling extolled the health benefits of ingesting large doses of vitamin C, an idea quickly snapped up by others to promote the belief that antioxidants can slow down the aging process. Today, news reports suggest that more than 500 milligrams of vitamin C per day may damage the genetic material of our chromosomes (DNA) rather than protect it. It is no wonder that some people are confused and ignore all health advice while others believe everything they hear or read. Scarcely a week goes by without some new book or news report on the latest "scientifically proven" way to extend life, reverse aging, or prevent heart disease and cancer. Extensive marketing campaigns attempt to convince baby boomers and the elderly that the secret to combating disease, maintaining health, retaining youth, and extending longevity has been discovered. Fads that were once restricted to health-food stores have mushroomed into a multi-billion-dollar industry that taps into the human obsession with aging, health, and disease.

The life extension industry begins with a grain of truth but quickly gets mixed with a tablespoon of bad science, a cup of greed, a pint of exaggeration, and a gallon of human desire for a longer, healthier life—a recipe for false hope, broken promises, and unfulfilled dreams. The quest for immortality has produced countless chefs who have cooked up fantastic stories about aging, and people are consuming their products in massive quantities.

Our goal in writing this book is to help you understand how and why aging occurs so that you can make informed decisions about your health, longevity, and quality of life. Thanks to modern-day technology, the proponents of extreme longevity spread their slick messages more widely than ever before. The bad news is that these new messages contain the same false promises that have been marketed and sold for thousands of years. The good news is that falsehoods, deceptions, and exaggerations are unnecessary. Scientists are on the threshold of discoveries about aging that are likely to have consequences for personal health and longevity that we could only have dreamed of just a few decades ago. We are optimistic that aspects of the aging process will eventually fall within the control of the biomedical sciences—permitting humanity, for the first time, an opportunity to influence the biological forces that govern life and death.

In this book, we offer a vision of human aging and longevity that is more realistic and far more promising than the false images and messages about aging that are so prevalent today. If we tell the story of aging correctly, then this book should give you a much better understanding of how and why we age, an expanded awareness of disease and death, and a greater appreciation of health and longevity. You will certainly be awed and maybe even frightened by the advances in biomedical technology that are going to change the meaning of life and death as they are currently understood. Perhaps more important, we hope that you will learn from this book that aging is far more than a biological process of decline and decay that culminates in death. Growing older can and should be a rewarding physical and emotional journey. For most of us throughout the majority of our lives, the passage of time can and should be marked by good health, improvement in physical fitness, emotional growth, and enhanced wisdom.

For over a decade, we have been conducting scientific research on individual, societal, and population consequences of aging. We have helped develop the emerging field of science known as the biodemography of aging, are recognized experts in population aging and the comparison and prediction of mortality between species, have participated in public policy debates about forecasts of human longevity and the solvency of the Social Security trust funds, and have advised life insurance companies in the United States and abroad. After years of publishing scientific papers, we believe the time has come for us to create something other than publications only scientists can understand. We felt an obligation to translate our own research on aging, and that of other scientists whose work sheds light on this topic, into something that can entertain, enlighten, influence, and hopefully benefit people outside the ivory towers of academia. We hope that you enjoy reading this book as much as we enjoyed writing it; we also hope that we have managed to convey the enthusiasm for research on aging that has made our work an avocation as well as an occupation.

Acknowledgments

—✕—

THE AGING PROCESS is complex and pervasive in its influence—affecting our bodies, our minds, our families, and society. In order to provide a thorough presentation of aging and its implications for the world within which we live, information must be gathered from the biological, medical, and social sciences, as well as from religious sources. Many people helped us along the way. We take this opportunity to thank these people and apologize to those whose contributions we fail to mention. Leonard Hayflick deserves special recognition. He took the time to review the entire book, provided encouragement and guidance, and loaned us his most precious books on the history of thinking about aging. Special thanks also go to Douglas Grahn and Michael (R.J.M.) Fry, mentors who provided honest critiques throughout the creation of this book and frequently reminded us to do our homework and check our assumptions. Dr. Andrus Viidik and Monika Skalicky reviewed the entire book and provided valuable comments that led to a number of important changes. Dr. Sherwin Neuland has been supportive from the beginning, providing advice and encouragement that was important in the early phase of our work. We would particularly like to thank Gayle Woloschak for her patience and lucid explanations for the linkages between molecular events and their impact on health and disease. Similarly, David Grdina was unfailingly helpful in explaining free radicals and the repair of the biological damage they cause. We would also like to thank Fred Stevens for his insights on structural biology and his visions of biomedical interventions in the future.

Dr. Randy Nesse provided valuable insights on evolutionary medicine and was kind enough to review several sections of the book, while Paul Eggers provided detailed statistics about the Social Security program at its inception early in the twentieth century. A special thank-you goes to Marilyn Webb, who removed the shades on our eyes about the process of dying, and whose advice and friendship has always been in abundance. On the religious side, we are in debt to our friends whose lives are devoted to the study of Torah and the Bible. Rabbi Yisrael Koval, Rabbi Avie Shapiro, Rabbi Stephen Hart, and Reverend Timothy Ek collectively spent an enormous amount of their personal time teaching us the nuances of the ancient scriptures. They are true scholars and it was a privilege to have them as teachers. Perhaps more important, they patiently endured our endless and politically incorrect queries while retaining a sense of humor.

Chapter 1 was based, in part, on the work of Gerald Gruman—a scientist who dove deeply into the morass of articles and books on the longevity movement dating back to the dawn of the written word. Although we read the original ancient texts when available, it was often the case that we relied on Gruman's scholarship and attention to detail. We gratefully acknowledge and recommend his work to anyone interested in the history of aging. It is a basic truth of scholarly pursuit that we all stand on the shoulders of those who have gone before us. We have constantly been amazed and humbled by their remarkable prescience and accurate conjectures on issues that had not yet been illuminated by the light of science.

It was our goal from the beginning to write this book for the general public. Initially, this was difficult because we were used to working in the densely coded language of science. Translating this language into something that can be understood and used by those not familiar with scientific terminology is, in part, why the public is inundated with misinformation on the aging process and what can be done about it. Joel Cohen advised us that as we drafted the chapters, we should pretend that we were talking face-to-face with someone who was not only unfamiliar with the science of aging, but also unfamiliar with science in general. We tried to follow his advice, but as a backup we relied on a number of people who read through or listened to various parts of the book in order to make sure that the language was clear and the messages strong. For this effort we are indebted to Linda Carnes, Michael Carnes, Tricia Dockery, Astrid Fletcher, Marsha Freeman, Tom Fritz, Ann Grahn, Ira and Beverly Hirsch, Lisa and Steve Kaufman, Muriel Magnin, Mike Miller, Abe and Pearl Olshansky, Jessica Olshansky, Sara Olshansky, Randi Schneider, Arlene Schultz, Jeff Schwartz, Tom Seed, Norm Slade, Len Smith, Sanna Stein, Jonathan Trent, Ben and Sylvia Turkin, Roberta and Mitchell Weisner, and Dolly and Armand Zucker.

We would not have been able to make a career out of pursuing answers to our own questions without funding. Over the years, we have received and in some cases still receive funding from the Department of Energy (DOE), the National Aeronautics and Space Administration (NASA), the National Institutes of

Health/National Institute on Aging (NIA), and the Social Security Administration (SSA). The SSA deserves special credit for having the foresight to support research on how the biology of aging affects attempts to forecast the number and age distribution of future beneficiaries. Our ongoing Independent Scientist Awards from NIA are the core support for our efforts to meld together the disciplines of biology and demography in order to study the biodemography of aging. We would particularly like to acknowledge the cooperative and supportive relationships that we have had over the years with funding officers from these governmental funding agencies—Marv Frazier, Mike Ginevan, Bob Goldsmith, and Bob Thomas at DOE; Frank Cuccinotta at NASA; Rose Marie Li, Georgeanne Patmois, and Richard Suzman at NIA; and Martynis Ycas from the SSA.

Our agent, Jim Levine, has been a source of support from the beginning. He not only helped shape our original proposal but also spent considerable time providing valuable comments on the content of the book itself. We wish Jim luck in his effort to apply our personal recipe for health and longevity in the last chapter. Our editors at Norton, Sarah Stewart and Drake McFeely, patiently and methodically shaped and molded the book into its current form. We are most grateful for their sound and sometimes painful advice, which led us to modify our language, remove some chapters and relocate others, and focus on the big picture. Our copyeditor, Ann Adelman, did a remarkable job of catching a number of errors and inconsistencies. A special thank-you goes to Mike Snell, who encouraged us early on to carry this work to its logical conclusion.

ON A MORE PERSONAL NOTE, I owe a great debt of gratitude to my immediate family—Sara, Jessie, and Ricky. They sacrificed countless weekends and evenings for three full years, leaving me alone to work on the book during times that should have been spent enjoying the pleasures of family life. Their love and patience has been an extraordinary gift. To my parents, Abe and Pearl, who wake up every morning thankful for another day, filling each with as much fun and enjoyment as is humanly possible. They taught me the importance of living one day at a time, a central theme in this book. My sister, Arlene Schultz, patiently read through the manuscript and counseled me repeatedly on how to communicate in a nonscientific language. Her advice was instrumental in helping us "talk" to our readers. I would also like to acknowledge a valuable contribution made by fellow fitness buffs at the Highland Park Hospital Health and Fitness Center in Buffalo Grove, Illinois, where portions of this book were written. I was inspired by watching people of all ages at every conceivable fitness level working out every day to improve their health and quality of life. They are all living proof that exercise is the only true Fountain of Youth that exists today. And finally, I want to thank my in-laws Ben and Sylvia Turkin for allowing us to begin the book with a true story about their efforts to engage in a battle that has occupied the minds of billions before them.

[S.J.O.]

LIKE JAY, I WOULD LIKE to thank my family for the personal sacrifices they made to turn this book into a reality: abbreviated and missed vacations, tying up the home PC that the kids wanted to use, missing their exploits on the soccer field, dumping most of my family chores on my already overworked wife, and last but not least, tolerating an escalation of my already cranky nature. I have one heartfelt regret that may fade but will never heal. My mom and dad knew about the book, but neither of them lived long enough to see it in book-stores, to hold it in their hands, and to proudly show friends what their boy did. I learned from my wife and children how not to be selfish. Almost everything else I learned from the example of my parents' total commitment and freely given dedication to family. Mom and Dad, I will never forget that parental love can never be repaid, it can only be passed on—a simple but profound truth passed on to me by the Reverend Paul Duncan. Thanks, Paul, wherever you are.

[B.A.C.]

The QUEST for IMMORTALITY

———⋈———

CHAPTER 1

——⋈——

Death and Immortality:
Early Views

Of all the questions which, throughout the centuries,
have escaped from the lips of man, there is none which
has been asked with such persistence than [sic]
"Where do I come? Whither shall I go?"

ANNIE BESANT

IN THE LATE NINETEENTH CENTURY, the French physiologist Charles-Edouard Brown-Séquard announced a miraculous discovery: the secret to rejuvenation. He removed and crushed the testicles of domesticated animals, extracted "vital" substances from them, then used the resulting concoction to inoculate older people against the "aging disease." His treatment appeared to work. Those inoculated reported improved mental acuity and physical vigor. When Brown-Séquard injected himself with this extract at the age of seventy-two, he claimed to have better control over his bladder and bowels.

Eugen Steinach, a professor of physiology in Vienna in the 1920s, found fame and fortune by convincing older men that they would be rejuvenated by a vasectomy or by having the testicles of younger men grafted onto their own. Soon, rejuvenation clinics sprang up everywhere and enterprising surgeons devised a number of antiaging therapies, including the application of electricity to the testicles and "stimulating" doses of X rays and radium to the sex organs.

Today, rejuvenation clinics throughout the world are employing modern versions of these bizarre methods. For example, at a rejuvenation clinic in the United States, a combination of hormones and antioxidants is ingested by and injected into people who have been led to believe that these "treatments" will slow down or reverse the aging process. A popular antiaging therapy in Europe is to journey into mountain mines where the elevated temperature, high humidity, and radon gas are believed to have therapeutic properties. For thousands of years, mankind has sought ways to halt or reverse aging, cheat death, and achieve immortality.

Death is an uncomfortable and often frightening subject. What most of us know about death we have learned from the movies. The feigned and often violent deaths portrayed in movies are abrupt terminations of life that rarely have anything to do with the process of aging. In the real world, terminal illnesses are often protracted, and death, when it comes, is so incomprehensible to those who witness it that it seems unreal.

Despite intense scrutiny, aging and death remain as two of humanity's great unsolved mysteries. In a world of high-tech medicine and extensive knowledge of human biology, aging and death continue to evade comprehension just as they have throughout a history dominated by magic, superstition, and pseudoscience. Perceptions of aging and death have changed dramatically over the course of history, influencing everything from religion and philosophy to science and medicine. What has remained constant is the effort to explain aging and death, and to find ways to escape them.

EARLY FATALISTS

The early history of religion, science, and philosophy was infused with an acceptance of aging and death as an inevitable and perhaps even desirable end to earthly life. Our ancestors were nearly defenseless in the face of a relentless onslaught of disease. They responded to their vulnerability by developing concepts such as life after death to soften its harsh reality.

From within this fatalistic view of aging and death sprang forth an idealized notion of what life could and perhaps should be. The tragedy of losing children to infectious diseases, young women during childbirth, and the eventual loss of old people to aging-related diseases, has led nearly every culture to create an idealized world within its written and oral folklore. In these utopian worlds, life exists without disease, disability, or death. The pleasures and vigor of youth are not only maintained indefinitely but are greatly enhanced by the wisdom that comes with increasing age. If human beings must struggle through pain and anguish just to live a few decades, then it was believed that there must be forms of life that benefit from our struggles, or places on earth where paradise reigns. There must be gods and a handful of lucky mortals who are not only free from disease but throughout their immortal lives remain forever young. These utopian worlds appear repeatedly throughout history as hopeful images of what life must be like somewhere—either on earth for mortals or in heaven for the deceased or the gods.

When our ancestors first wrote about such utopian worlds,

they did not settle for the mere elimination of just one plague or disease. They wanted it all—eternal youth and immortality. For those who had already experienced the ravages of time, they wanted rejuvenation. For their loved ones who had already died, they wanted resurrection. These idealized worlds, created to soften the reality of life in a world filled with pain and suffering, continue to play a prominent role in the myth, legend, and folklore of aging and death today. The unending quest for immortality has transcended geography, culture, and time—a quest that has probably touched not only your life but the lives of most people in modern times.

The Greek legend of Prometheus and Pandora, from the eighth or ninth century B.C., provides one of the first mythical accounts of the origin of death. Prometheus, known for his cunning and favor with mortals, was a son of Iapetus, the Titan who was overthrown by the god Zeus. When Prometheus aroused the wrath of Zeus by offending him, Zeus punished Prometheus and the humans to whom he was sympathetic by barring fire from mortals. In defiance, Prometheus stole the fire and carried it down to earth. In retaliation for this act, Zeus decided to "give men as the price for fire an evil thing in which they may all be glad of heart while they embrace their own destruction." Zeus created Pandora, the first woman, who had "charming traits" to hide her wicked nature. The jar that Pandora brought with her to the world was filled with evil things, including among them plagues, disease, and old age. The Prometheus legend was intended to illustrate not only the great power of Zeus, but also that old age and death are due to the will of the gods and are thus unchangeable. The idea that women are responsible for aging and death is a recurring theme in a history chronicled principally by men.

The Babylonian legend about Gilgamesh, who was part-man, part-god, is another lesson on the inevitability of aging and death. Gilgamesh was an arrogant young king who treated his subjects poorly. In an attempt to teach the king humility, the gods created Enkidu—a man of incredible strength who promptly engaged the young Gilgamesh in combat. Combat eventually gave way to mutual admiration; they became the best of friends and traveled the world seeking adventure. But after they had inappropriately directed their bravado at a goddess, the gods killed Enkidu—leading Gilgamesh to contemplate his own mortality.

Obsessed with obtaining immortality, Gilgamesh sought the advice of a sage. He was first told that immortality could be achieved by mastering sleep and was instructed to stay awake for six days and seven nights. Weary from his long travels, he eventually fell asleep. His last hope was to obtain a plant at the bottom of the sea that was believed to possess the power of rejuvenation. After obtaining the plant, Gilgamesh took time to bathe in a pool of cold water. As he bathed, a serpent appeared and ate the plant—providing an early rationale for why snakes appear to renew their lives by shedding their old skins. The moral of this legend was that if death was inevitable for a part-god with Herculean strength, then ordinary people must also accept the fate of their mortal lives.

The story of Adam and Eve provides a biblical explanation for the origin of death. Upon being placed in the Garden of Eden, God told Adam that he could eat fruit from any tree except the Tree of Knowledge. He could gain immortality by eating fruit from the Tree of Life, but death would be the penalty if he ate from the Tree of Knowledge. Eve, seduced by the serpent to eat from the Tree of Knowledge, enticed Adam to do the same. The consequence was that humans became subject to death: "in the sweat of your face you shall eat bread till you return to the ground, for out of it you were taken; you are dust, and to dust you shall return" (Genesis 3:19; Revised Standard Version and other translations for subsequent quotes).

Epicurus, the fourth-century B.C. Greek philosopher, carried the fatalistic view a step further. He proposed that a tranquil life could be attained by removing the fear of death through a proper attitude. The Epicureans also believed in the "fullness of pleasure," a concept derived from the premise that it was futile to live forever because life offered only a limited number of gratifications. An oversimplified version of this concept goes something like, eat, drink, and be merry, for tomorrow we die—a modified passage from Isaiah in the Old Testament. The Roman poet Lucretius in the first century B.C. also argued that it was pointless to prolong life because no matter how long you live, the length of the time you spend alive is insignificant compared with the infinite time you spend dead.

In the second century A.D. while Marcus Aurelius was emperor, Stoicism was the official philosophy of the Roman Empire. Average length of life then was roughly 25 years, which implies that approximately one-third to one-half of all the babies born at that time died before reaching their first birthday. Contemporary philosophers tried to rationalize this cruel reality by suggesting that length of life then was roughly a matter of no importance, and that death was a fact of life that everybody must accept. The Stoics even went so far as to suggest that death should be embraced because it was natural and necessary for the proper functioning of the universe. In a world where infectious diseases routinely decimated the young, it is not surprising that the Stoics developed a philosophy that not only rejected the extension of life but embraced death.

The EARLY BIOLOGY *of* FATALISM

A biological perspective on old age and death first appeared in essays written by the Hippocratic physicians of the fifth century B.C. and epitomized by the work of Aristotle. The Hippocratic physicians divided the lifespan of mankind into four "regions," each with its own unique qualities associated with aging. Childhood was hot and moist; youth was hot and dry; adulthood was cold and dry; and old age was cold and moist. Aristotle modified this scheme by merging adulthood and old age into a single region that he described as cold and dry. Aristotle believed that aging and death came about by a transformation of the body from one that was hot and moist to one that was cold and dry—a change that he believed not only inevitable but also desirable.

Aristotle's cosmology was based on fundamental differences between the physical components of earth and heaven. He believed that all things on earth, including living things, were composed of four basic elements: earth, air, fire, and water. Everything in the heavens (sun, planets, and stars) contained the four basic elements plus an additional one called ether. Only things containing ether remained unchanged through time. Immortality was a state of existence restricted to the heavens. On earth, aging and death were considered by Aristotle to be the natural consequences of inevitable change and decay. Aristotle admired nature's ingenuity because everything appeared to occur in a well-planned and organized way. For example, he noticed that teeth fall out in old age because the nearness of death makes them no longer necessary.

Two other intellectuals whose ideas about old age and death continue to influence modern thinking about aging were the Greek physician Galen from the second century A.D.

and the Arabic philosopher/physician Avicenna from the tenth to eleventh century A.D. Galen believed that aging began at conception. He believed that heat from the male sperm began a drying process that initially stimulated growth and development, but by early adulthood, the drying process shifted from a beneficial to a harmful mode. This transition was characterized by a loss of innate moisture—eventually leading the body to become cold and dry. Although immortality could theoretically be achieved by retaining the innate moisture of the body, Galen believed the drying process was not only inevitable but a natural part of the order of the universe. According to Galen, aging was not a disease but a natural and expected phenomenon.

The views of Avicenna from the Islamic tradition were similar to those of Aristotle and Galen. Besides believing that aging was inevitable, Avicenna had the remarkable insight to suggest that "the art of maintaining the health is not the art of averting death . . . or of securing the utmost longevity possible . . . the art of maintaining health consists in guiding the body to its natural span of life. . . . " Even a thousand years ago, insightful scholars like Avicenna knew that life should not be lived as a constant battle against death, but rather as the daily pursuit of a healthy life.

The RELIGIOUS LEGACY

The Old and New Testaments of the Bible have had an important influence on the philosophical concept of fatalism (the belief that aging and death are not only inevitable but desirable) as well as on scientific hypotheses about old age and death. In the Old Testament, a period of human history is described during which the ancient patriarchs supposedly lived much longer than subsequent generations. Naturally, there is considerable debate about whether these reported ages are a metaphor or should be taken literally. Adam supposedly lived 930 years and was 130 years old when his third son, Seth, was born. Methuselah, who lived 969 years, holds the record as the longest-lived patriarch, while Noah, who was reported to have lived 950 years, was the last of the extremely long-lived patriarchs. Just before the Great Flood, God said: "My Spirit will not contend [remain in] man forever, for he is mortal; his days will be a hundred and twenty years" (Genesis 6:3). This biblical reference to a 120-year lifespan for mankind is often erroneously presented as scientific fact.

The Old Testament states clearly that God is in complete control over aging, disease, and the longevity of every living thing. For example, from Ecclesiastes (3:19), "Man's fate is like that of the animals; the same fate awaits them both: As one dies, so dies the other. All have the same breath [spirit]; man has no advantage over the animal." From Job (14:5), "Man's days are determined; you have decreed the number of his months and have set limits he cannot exceed." And from Proverbs (10:27), "The fear of the Lord adds length to life, but the years of the wicked are cut short." In Isaiah (65:20), God promised the Babylonian exiles who returned to Jerusalem that he would give them health and longevity in their new kingdom: "Never again will there be in it an infant who lives but a few days, or an old man who does not live out his years; he who dies at a hundred will be thought a mere youth; he who fails to reach a hundred will be considered accursed."

A significant transition occurs in the New Testament. Long life is no longer offered as a reward for righteousness; instead, there is almost an indifference to matters of the body. While the Old Testament makes it clear that the original sin of Adam is responsible for the judgment that brought death to mankind, the New Testament decrees that eternal life is accessible to everyone because of the sacrifice that God made of His only son: " . . . by the trespass of one man [Adam], death reigned through that one man . . . so also the result of one act of righteousness was justification that brings life for all men . . . eternal life through Jesus Christ. . . . " (Romans: 5:15–21). The New Testament promises spiritual immortality through death and subsequent resurrection. Death becomes, once again, not only necessary but desirable.

Some of the most interesting ideas on fatalism come from the writings of the thirteenth-century mystic St. Thomas Aquinas and of the earlier Christian father St. Augustine. According to St. Thomas, before Adam ate from the Tree of Knowledge, his body was subject to the ravages of time. However, God gave him a supernatural power to combat the natural deterioration experienced by other living things. After eating the apple from the Tree of Knowledge, God withdrew this supernatural power. Thereafter, all humans were subject to the natural process of deterioration that led to old age and death. Interestingly, Aquinas linked death to the inability of men and women to control their minds or maintain their willpower. The implication is that by reasserting control over the mind, the supernatural power taken by God could be restored and perfect health and immortality achieved. This mind-over-body philosophy remains an important element of the ideology that drives the modern antiaging movement.

For his part, Augustine of Hippo believed that the loss of mind-body control not only leads to death but also to sexual lust. This sexual tension and linkage between sex and death can be found throughout the Old and New Testaments. In a broader sense, this linkage is also a common thread that weaves its way through the extensive historical literature on longevity.

PROLONGEVITY

Not everyone shared the resignation of the fatalists. Prolongevists, as they have come to be known, believe that aging and death are amenable to modification, and that longevity can be extended through human intervention—a belief that has persisted throughout history. Although all prolongevists believe that longevity can be extended, their techniques to achieve this goal have differed dramatically. Some prolongevists have claimed that a few years can be added to life, while the more radical prolongevists have claimed that the elimination of human disease is going to make the dream of eternal youth and immortality a reality.

Taoists were the first to develop a systematic effort to prolong the lifespan. Taoism is a religion formed in ancient China during the third century B.C. Its ideology and strict regimens for living are based on a fundamental belief that the prolongation of life is not only possible but highly desirable.

At the center of Taoist philosophy is the concept of *tao* (pronounced *dow*), which translates into "the way" but has been interpreted to mean "the mother of all things" or "the unity of nature." Unlike the Greek philosophers, the ancient Chinese did not strictly separate spirit and matter—making it possible and even desirable to experience a transformation from the physical body into an immortal spiritual being. An early and influential Taoist thinker, Ko Hung, suggested that animals could be changed from one species to another, that lead could be transformed into gold (the basis of alchemy), and that mortal humans should strive to become immortal beings (known as *hsien*). The legends about *hsien* gave birth to the Taoist methods for extending the lives of mortals.

There are many prolongevity elements in Taoist thinking,

but perhaps the most important are the concepts of quietism and primitivism. Quietism is a philosophy of life and a way of living that is captured by the term *wu wei*, which translates to "effortless action." Effortless action involves a lifestyle and way of thinking that dampens the emotions. Taoists believed that people are endowed from birth with a fixed amount of "vital breath" that is consumed like fuel by emotions that come from the heart. A more passive life without honors and riches would conserve vital breath and thus permit the person to live longer. With proper discipline, the use of vital breath could be stopped altogether and the person would be trans- formed from a mortal being to a *hsien*. The ancient Chinese considered the "true men of old" to be those who perfected quietism and embraced primitivism—devoting themselves to a simple and modest lifestyle.

Taoists developed dietary practices in order to starve and drug "evil beings"—referred to as the Three Worms—who were thought to inhabit the body and hasten its demise by causing physical problems and disease. Battling the evil beings took the form of denying them the grains (such as wheat and rice) thought to be responsible for their existence, and eating magi- cal foods like cinnamon, licorice, cinnabar, and ginseng that would kill them. Other *hsien* medicines included herbs, roots, and minerals, and animal and plant products such as eggs, tur- tles, peaches, and parts of trees. All were revered for their abil- ity to enhance longevity. Some of these products are still being sold and promoted today. A peach-flavored tea referred to as Longevity Tea is popular in many restaurants in the United States. Many modern myths about remedies that forestall or reverse aging are descendants of these ancient Taoist beliefs.

Central to early Taoist thinking was a linkage between the mortal body and the immortal heavens. Taoists envisioned the air entering the lungs as blending with the sky—the same sky that reached up to merge with the heavens. By controlling the breath, a more intimate physical contact with the heavens could be achieved. If breathing techniques could be perfected, then immortality could be attained. A series of breathing exercises was devised to achieve the ultimate goal of holding one's breath for the time normally required for 1,000 respirations. Techniques were also devised to promote the conscious guidance of inhaled air through various parts of the body, and through a method referred to as "embryonic respiration" to nourish the body by extracting nutrients from the air rather than food. Adherents of these practices learned how to dramatically reduce their metabolic rate and use of oxygen; they also adopted a near-starvation diet consisting mostly of roots, berries, and other fruits. (There are scientific studies today that appear to support the benefits of the dietary practices of Taoists.)

According to Gerald Gruman, a scholar of prolongevity, "Taoism's greatest achievement was to take prolongevitism from the realm of magic and carry it forward to a stage which is best termed proto-science. Although Taoists did not originate the notion of prolongevity; what they did was to take the prolongevitist vagaries of folklore and form them into an organized body of concepts and hypotheses suitable for assimilation by science." In addition, the Taoists provided the first systematic effort to identify specific foods, medicines, and chemicals associated with aging and disease. They also proposed behavioral changes that form a conceptual basis for several scientific and pseudoscientific efforts currently being promoted as ways to slow down or reverse the process of aging. Taoist exercises associated with longevity have been an influential force throughout the centuries and are still practiced today in the form of Tai Chi Chuan, acupuncture, and techniques of what is now known as Swedish massage.

The ALCHEMY *of* PROLONGEVITY

Alchemy grew out of the ancient belief that ordinary metals could be transmuted into silver and gold. Considered by most to be a noble profession, alchemists were revered members of their community. According to Dr. John Read from The University in Scotland, who chronicled the historical roots of this ancient form of chemistry in his book *From Alchemy to Chemistry* (1995), the great alchemists thought of their work as a sacred trust. Famous alchemists stood beside rulers and kings, commanded great sums of money, and lived privileged lives. Some of the more unsavory ones used chemistry that would be rudimentary by today's standards in order to dazzle investors and an unsuspecting public into believing that they would eventually be able to transform elementary metals into gold.

Like any good scientist today, the alchemists of the early prolongevity movement were systematic and methodical. Immortality was simply a puzzle; it was only necessary to find the right pieces and put them together in the right order. Alchemists believed that a Fountain of Youth existed, and their investors backed this belief with massive sums of money. Through an orderly line of reasoning and rigorous scientific investigation, the alchemists believed that all the proper steps required to transform mortal men into immortal beings would eventually be discovered.

The earliest applications of alchemy involved efforts by the Chinese to cure disease and extend longevity. Ko Hung, the Taoist, was one of the first outspoken proponents of pro-longevity. He considered arrogant and dogmatic the prevailing attitude that death was inevitable and immortality impossible, especially given the limited knowledge about aging and death that existed at his time. The metamorphosis of caterpillars into butterflies, tadpoles into frogs, and seeds into flowers gave visible proof from nature that the transformation of living things was possible. If metals and living things could be transformed from one thing to another, then transformations from sickness to health, and from mortality to immortality, should also be possible.

To achieve these transformations, Ko Hung believed it was necessary to find and purify highly vitalized foods—herbs, minerals, chemicals, and other substances that, when ingested, would increase the "life force." These life-sustaining substances had visible properties which he described in detail—shining or bright, fluid or wet, sharp or bitter in taste, red or scarlet in color, and often shaped like a man or animal (root plants, for instance). Gold was an especially important *hsien* substance because it was resistant to chemical change or, as the Taoists would say, noncorruptible in its nature. As such, it was a highly desired substance that was thought to greatly enhance longevity. Several methods were devised for ingesting gold, including mixing it with organic materials to swallow in pill form, or by making cups of gold from which one would drink and so ingest minute quantities of the valuable substance. What is particularly fascinating about gold was that it was often desired not because of its enduring beauty or monetary value, but because it was considered the most potent antiaging substance known to mankind.

Another important substance thought to promote longevity was cinnabar—known today as mercuric sulfide. This substance was coveted because it was blood red in color and turns into a silvery and slippery liquid (mercury) when heated—both properties associated with longevity. Mercury, in turn, transforms into a red powder known as mercuric oxide when heated. According to the Chinese alchemists, any substance maintaining the visible properties associated with longevity, when challenged by heat or organic chemicals, would confer enhanced longevity if consumed in small quantities.

The use of alchemy to influence aging and disease eventually spread from the Chinese to the Arab world. Based on the extensive writings on this subject by the Arab physician Jabir, this flow of information must have occurred by the eighth century A.D. It appears that Jabir was primarily interested in alchemy for its medicinal purposes—creating elixirs to treat and cure disease. He also espoused the belief that men could master the forces of nature, a basic tenet of the prolongevity movement. It was through Jabir that Taoist thinking about aging and disease was transmitted to the West. After Jabir, many historical figures were important to the prolongevity movement. But two in particular have had a profound influence on modern views about methods of extending life: Roger Bacon and Luigi Cornaro.

PROMINENT PROLONGEVISTS

Many consider Roger Bacon, an English philosopher and scientist of the thirteenth century, the father of the modern prolongevity movement. He believed that there was no fixed limit to life and that the shortened lives of his contemporaries, relative to the biblical patriarchs, were due entirely to immoral and unhealthy lifestyles. He also believed that life could be extended by using the "secret arts" of the past, which to him involved the use of life-prolonging chemicals, foods, and other substances. Like Galen and Avicenna before him, Bacon believed that old age was due to the loss of innate moisture with the passage of time.

Although these ideas were not new, Bacon did contribute a

new twist to the philosophy that nature could be controlled and longevity extended. He suggested that the negative effects on longevity caused by immoral and unhealthy lifestyles were not only passed from one generation to the next, but they were cumulative over time. To Bacon, this explained the progressive decline in the length of life that had occurred from the time of the ancient patriarchs. According to Bacon, lifespans would continue to shorten every generation even if there was no change in behavior from the preceding generation. However, Bacon also believed that this process could be reversed and the human lifespan restored to the 900 years or more lived by the ancient patriarchs. His solution to increasing longevity comes straight out of the Fountain of Youth tradition that is still popular today. Substances like pearl, coral, aloe wood, gold, and "bone from a stag's heart" were thought to contain an innate moisture or vital spirit which, when ingested in small quantities, would replenish that lost during the course of life.

Bacon and his contemporaries also recommended the breath of a young virgin as a form of antiaging therapy for old men. Since the young were thought to possess a large supply of the "vital principle," or breath, it was believed that older men could absorb some of it by being in their company. By the thirteenth century it was already known that breathing the same air as someone who was sick could spread disease; it seemed plausible, therefore, that health could be transmitted in the same way. This form of therapy for old age became quite popular. Curiously, breathing the air of young virgin boys was never mentioned as an antiaging therapy for older women.

Another approach involved what has become known as the Philosopher's Stone. Bacon and the alchemists who preceded him all believed that the transmutation of metals and living things was possible. One of the great dreams of every alchemist was to transform gold into a purer form that would have even more powerful antiaging properties than regular gold. The idea that gold could be made purer led to the belief that all things had a more pure essence. The purer form could be attained through the use of the Philosopher's Stone, which was a mythical stone (or powder) of great power that was thought to be created through a series of chemical transformations using a secret formula involving the techniques of alchemy—heating, distilling, and dissolving. According to the alchemists, these purified substances held the most promise for altering the human lifespan. The contemporary search for biochemical means of lengthening the lifespan has its origins in the attempts made by alchemists to extend life through transmutation.

Luigi Cornaro, a member of the Italian nobility in the fifteenth century, also had a profound influence on the modern prolongevity movement. For the first forty years of his life, Cornaro admits that he pursued an indulgent lifestyle replete with extravagant food and "frequent overindulgence in sensual pleasure." When his health began to fail, a physician warned Cornaro that his self-indulgent habits would lead to his early demise. Once the diet prescribed by his physicians was modified to suit his own tastes, Cornaro experienced a complete recovery from his illnesses. Cornaro went on to live a healthy and happy life, free from disease, and greatly enjoyed the company of his eleven grandchildren right up to his death at the age of ninety-eight.

Cornaro offered a simple approach to healthful ways of living, which explains why it has been accepted and heartily endorsed for centuries after his death. He accepted the common belief that everyone is born with a fixed quantity of "vital principle" required for life, which is used up during the course of living. Cornaro also believed that even the weakest people are born with enough "vital principle" to live 100 years; those endowed with a stronger constitution could live to the biblical maximum of 120 years. Given contemporary rumors of people living to 150 years or more, Cornaro believed that even these incredible ages could be achieved.

His recipe for a healthful lifestyle was to eat foods that agreed with you, and to eat them in quantities that the stomach could digest easily. For Cornaro, this meant small portions of bread, meat, broth with egg, and light wine. He emphasized, however, that this regimen would differ for each person depending on their personal desires. Cornaro also suggested that with advancing age, the amount of food ingested should be reduced because the "natural heat" of the elderly decreased, and thus less fuel was needed. Avoiding the extremes of life, such as excessive heat, cold, fatigue, and hatred, was also important.

The central element of Cornaro's philosophy was that by nature man is healthy. From his perspective, all diseases were caused by unhealthy lifestyles. Cornaro carried his reasoning one logical but big step further when he suggested that all diseases could be avoided. When scientists today make predictions of extreme life expectancies, their assumptions bear an uncanny resemblance to those of Luigi Cornaro.

The THREE BRANCHES *of* PROLONGEVITY

Three distinct themes characterize the myths, legends, and pseudoscience of prolongevity. One is based on what are known as Antediluvian legends—the belief that people from the distant past lived far longer than people today. Hyperborean legends comprise the second theme—the belief that there are places in the world where people live exceptionally long lives. The third and best known theme involves the Fountain legends—magical waters, foods, herbs, minerals, or other substances believed to have the power to rejuvenate the old, resuscitate the dead, bestow immortality among the living, and cure every disease known to afflict mankind.

The Antediluvian Legends

Several cultures and religious traditions maintain legends that our ancient ancestors lived much longer than people live today. According to Hebrew tradition, many of the patriarchs listed in the Bible lived extremely long lives. There are at least two biblical interpretations of the ages reported for these patriarchs. Those who interpret the Bible literally contend that there was a time in history when mankind was either immortal (in the Garden of Eden during the time of Adam) or extremely long-lived (during the time between Adam and Noah). Others suggest that the extreme longevity of biblical patriarchs should not be taken literally; instead, their long lives were symbolic of their importance in the history of mankind.

The Greek philosopher Hesiod, who gave the first account of the Prometheus legend, provided another Antediluvian theme. He divided the history of mankind into five distinct epochs. The first epoch was the Age of Gold—a time of peace and tranquility when mortals never aged. During the Age of Silver, childhood lasted a hundred years but adults lived only a short time. Hesiod's own epoch was the Age of Bronze—a time of deceit, injustice, and envy, when violence threatened to destroy the world. When the Bronze Age had extinguished, Zeus created a godlike race known as demigods who lived life without pain and suffering in what was referred to as the Heroic Age. The fifth age of mankind, the Age of Iron, is our world of today which is characterized by violence and the love of profit. According to Hesiod, Zeus will destroy this race of mortal men because we will dishonor our parents in their old age by not repaying them for the cost of their nurture. Hesiod explained old age and death and a decreasing lifespan as the product of a lifestyle that had become increasingly more decadent with time. Despite this dismal trend, Hesiod was optimistic that the forthcoming epoch might be better. The imagery of an idyllic distant time in the past when people were forever young, followed by a progressive trend toward shorter lifespans driven by increasingly more decadent lifestyles, is a classic Antediluvian theme.

The Hyperborean Legends

According to Greek mythology, there was a place on earth where people lived long lives free from disease and aging. These people were referred to as Hyperboreans—*hyper* meaning "beyond" and *Boreas* meaning "the North Wind"—because they were thought to occupy a place in the far northern hemisphere. Many distinct cultures and time periods possess mythologies about places of paradise, referred to collectively as the Hyperborean legends.

For much of human history, limited mobility and barriers to communication meant that most people spent their lives in relative isolation. These times and cultures gave birth to the Hyperborean legends. Early examples include reports of a 150-year lifespan for people living on Mount Tmolus in Asia Minor, and a 400-year lifespan for people living on Mount

Athos in Greece. The notion that paradise is not only real but is also accessible to everyone is a common theme among these early legends. Later, the major religions modified this theme to incorporate a paradise that was accessible only after death, and even then one that was restricted to those whose lives had been governed by the values of their religion.

A classic Hyperborean legend comes from India. In this legend, there existed a land far to the north called Uttarakurus where the people enjoyed perfect health, lived for a thousand years, and experienced dazzling sensual pleasures. The source of these wondrous benefits was the fruit of the Jambu tree, which conferred immunity from illness and old age. As is often the case with Hyperborean legends, it is unclear whether mere mortals had access to these lands or whether such places were accessible only to the gods or those who had already died.

Similar themes weave their way through Hyperborean legends throughout history. In the Middle Ages, countless ships set sail across the Atlantic ocean in search of legendary islands such as St. Brendan's Island, Avalon, Atlantis, and Antilia, thought to contain paradise. At the time of Columbus, it was widely believed that paradise was located on top of three mountains overlooking four rivers—the Tigris, Euphrates, Nile, and Ganges. In fact, Columbus was convinced that he had discovered the long-sought-after Terrestrial Paradise when, on his third voyage to the Americas, he spotted the confluence of four great rivers along the coast of Venezuela. Modern times have made their own contributions to Hyperborean legends: consider the image of Shangri-La portrayed in James Hilton's book *Lost Horizon* (1933), the quest for the Holy Grail in the movie *Indiana Jones and the Last Crusade* (1989), the pursuit of the Fountain of Youth in the movies *Cocoon* (1985) and *Cocoon: The Return* (1988), and the recurring stories of extreme longevity among people living in the Caucasus Mountains of Soviet Georgia.

The Fountain Legends

The most familiar of all the myths and legends about longevity are those based on foods, waters, and other substances believed to possess special powers to combat disease, aging, and death. The Hyperborean and Antediluvian legends often involve the use of one or more of these substances. Ponce de León is the best known of those who have searched for the Fountain of Youth. This Spanish explorer, who traveled with Columbus on his second voyage to America, established a colony in 1508 in the West Indies on the island now known as Puerto Rico. The local Indians told the Spanish explorers about an island called Bimini where there was believed to be a spring that restored youth to all who bathed in it or drank its magical waters. While searching for Bimini in 1513, Ponce de León accidentally discovered what is now Florida and the Bahamas. Legend has it that the fifty-five-year-old explorer was searching for the Fountain of Youth because he was no longer able to satisfy his young wife. It is also possible that he fell prey to an ingenious ploy by the local Indians to get rid of the unwanted Spanish explorers. Europeans took Ponce de León's exploits seriously, particularly after a high-level official testified to the existence of the Fountain and reported that the father of one of his servants had been revived by the spring waters.

Although the Fountain of Youth legend has become synonymous with Ponce de León, its origins can be traced back to at least 700 B.C. There is an old Hindu legend about an elderly priest named Cyavanna who married the daughter of the

Hindu king as part of an agreement to end a conflict. Subsequently, two demigods known as Asvins had become enamored with the king's daughter and tried unsuccessfully to seduce her. Cyavanna, who was aware of their efforts, offered to reveal religious secrets to the Asvins in exchange for rejuvenation. Accepting the bargain, the Asvins took Cyavanna to a Pool of Youth, where upon bathing, the three emerged to discover happily that they were youthful once again.

A Fountain legend also appears in the Old Testament, where there is a description of a River of Immortality that confers eternal life. Although specific references to a Fountain of Life or a river flowing out of Eden appear infrequently in the Bible, the concept of eternal life associated with these symbols is an integral part of Christian faith.

One of the more elaborate Fountain legends involves the quest by Alexander the Great to find the Fountain of Youth. According to this legend, a dried fish suddenly came back to life when it was being cleaned in a spring by one of Alexander's personal cooks. The cook promptly bathed himself in the water and became immortal. When the cook refused to reveal the location of the spring, Alexander tried to kill him. Discovering that an immortal could not be killed, Alexander threw him into the sea, where the cook continued to live in demonic form. In the Arabic version of this story, an angel tells Alexander about a well containing the Water of Life. Instead of a cook, it is a general and adviser to Alexander known as el Khidr who is transformed into an immortal after having accidentally discovered the well. A third version of the legend appeared in a twelfth-century poem. While on expedition near India, Alexander comes upon four aged men who tell him of three fountains: one restores the dead to life; the second bestows immortality; and the third has the power of rejuvenation. Upon finding the Fountain of Immortality, they discover that it can be used only once a year and someone has already beaten them to it. Alexander and the four aged men eventually locate the Fountain of Rejuvenation. After bathing in it, they happily discover that they have all returned to the youthful condition they had when thirty years old.

LEGENDARY QUESTS FOR IMMORTALITY and perpetual youth appear to spring from a near-universal fear of the aging process and its consequences. This fear of death and the unknown that follows it begins in early childhood and remains a concern right up to the moment of death. From the Greek story of Prometheus to the Holy Grail pursued in the Indiana Jones movie, the Antediluvian, Hyperborean, and Fountain legends about aging have played a prominent role in virtually every human society. Such myths and legends have been passed down through the ages and persist in our thinking today. However, as scientists begin to unlock the secrets of aging, an even more fascinating story begins to unfold.

CHAPTER 2

—⋈—

Sex and Death

The difference between sex and death is that
with death you can do it alone
and no one is going to make fun of you.

WOODY ALLEN

A FEW YEARS AGO, my wife and I were reading in bed late one night. I was enjoying the feeling of accomplishment that comes from sending off a completed manuscript to a scientific journal—the culmination of three years of research. Just before turning off the bedside lamp, Sara turned to me and asked about the manuscript. She has much more than a passing interest in our work—Sara and all of her friends are preoccupied with their own aging and how time has changed their bodies and minds, as well as those of their husbands. "Could you explain your latest research briefly, in a single sentence?" she asked. Although the full story is complicated, our work had established a link between when reproduction and death occur. "I can't explain this entire concept in a single sentence," I said. "Try anyway," she insisted. I paused for a moment, then blurted out, "The price we pay for sex is death." Without a moment's hesitation, she smiled at me and said, "Okay, then, no more sex."

Her answer caught me off guard. "Wait a minute," I said. "Let me tell you the full story of how sex, death, and immortality are related to each other."

Genes, Organic Time Travelers

The bodies of all creatures that reproduce sexually contain only two kinds of cells—*somatic cells* (the body you see reflected in the mirror) and *germ cells* (the eggs carried by females and the sperm carried by males). Germ cells contain a single copy of the master recipe for life—deoxyribonucleic acid (DNA). Somatic cells contain two copies of DNA, one from each parent. The first somatic cell of each new life forms when a sperm fertilizes an egg; at that moment, the engine of life switches on. In the process of creating a new life, the parent's DNA are shuffled like a deck of cards. The result is that the instructions for building and operating a human body, encoded within the DNA of our ancestors and passed through every subsequent generation, comes to us. We, in turn, reshuffle the genetic deck, and pass the DNA on to our children.

The diversity of life that so intrigues and entertains us is perpetuated every time a new somatic cell is formed. The DNA then guides the first somatic cell to produce more somatic cells; cells that ultimately form the tissues, organs, and bodies of living things like trees, beetles, birds, and humans. Human bodies, which can be referred to as *somas*, function like genetic transport vehicles. Genes have been hitching a ride in these vehicles through the ages. Neither can survive without the other. Human beings and other living things are directed to grow, develop, and reproduce so that the genes hitching a ride within us can move on to the next generation.

It follows that it is genes which are immortal, not the bodies that carry them. Genes, the ultimate time travelers, transcend the bounds of time that measures the limits imposed on our mortal bodies. As Richard Dawkins wrote in *River Out of Eden*: "A river of DNA . . . flows through time, not space. It is a river of information, not a river of bones and tissues: a river of abstract instructions for building bodies, not a river of solid bodies themselves. The information passes through bodies and affects them, but it is not affected by them on its way through."

IMMORTAL GENES, MORTAL BODIES

Scientists involved in the Human Genome Project have embarked on a quest to create a code book for the 80,000 to 100,000 active genes that make up human DNA. Imagine for a moment that you could read this genetic code book. You would find it to be an ancient scripture, written some 130,000 years ago. Most passages have remained unchanged from the original text. But like an evolving manuscript, some words and phrases would be crossed out and revised. Other than a few editorial changes here and there, the basic story in the human genetic code book has remained unchanged through time.

Now imagine a library full of the genetic code books for all life on earth, past and present. Most of the books would tell a similar story. In fact, some stories would be so much alike that an editor would suspect plagiarism. For example, humans and chimpanzees share more than 98 percent of their genetic story. Even species as different as humans and bacteria have genetic passages that are virtually identical. The consistent story told by these code books is powerful evidence that the history of life on earth has been, and continues to be, nothing more than variations on a single story of life that began some 3–4 billion years ago.

Having read the genetic story of our human ancestors, we can move on to the passages that describe our own lives. Like those who have gone before, one's personal genetic history chronicles a long and arduous journey. Countless obstacles and missed opportunities are revealed. First, there was the death of millions upon millions of sperm seeking an egg that had already been fertilized. Preceding you are countless fertilized eggs containing an incorrectly transcribed passage from the genetic code book—most of which were lost to spontaneous abortions. Numerous embryos that have developed abnormally have been reabsorbed into the wall of the uterus. The perils faced by a new life are never greater than at the time between fertilization and birth. Those of us alive are among the lucky ones, representing only a tiny fraction of fertilized eggs given the chance to pass on the ancient genetic scripture of life.

Once born, the journey through life is fraught with danger. Human beings are born with immature immune systems that leave infants and children mostly defenseless against the true rulers of the world—the bacteria and viruses that cause infectious diseases. Avoiding contact with these potentially deadly microscopic organisms is not possible. In our ancestors' time, the vast majority of people were killed by these organisms. Miraculously, we and all of our direct ancestors survived. The journey through childhood is no less perilous—potentially life-threatening accidents abound. Finally, we make it to a biologically critical moment in life, puberty.

The immortal book of life can now be passed on to yet

another generation. Once we reproduce, a new story begins, and so it goes through time. For us mortal somas, our story now becomes one of when and how nature discards us. But why are we discarded at all? Why doesn't the genetic recipe that we inherit from our parents build immortal bodies? Wouldn't an immortal body allow the story of life to continue just as easily as immortal genes? In a word, the answer to this question is a resounding no!

From a human perspective, the consequences of sexual reproduction—aging and the eventual death of the soma—seem cruel and wasteful. Even from a biological perspective, sexual reproduction seems terribly inefficient. When bacteria reproduce, they pass on their entire genetic code book intact. Sexually reproducing organisms, on the other hand, can only contribute half of their genetic code book into either eggs or sperm. A new life cannot start until the code book is made whole once again by the fertilization of an egg by a sperm. This means that our personal genetic recipe for life is diluted by one-half every time we reproduce. If passing genetic information on to the next generation is so important, why should something as inefficient as sexual reproduction even exist? When bacteria divide, they create duplicates of their genetic code book. The offspring of sexually reproducing organisms, on the other hand, receive a brand-new book that melds together recipes from both parents. The same old recipes work just fine when the world around you does not change. However, in a rapidly changing environment, the genetic diversity constantly created by sexual reproduction provides a much greater chance that at least some offspring will be produced with the right genetic recipe to survive and thrive. This provides humans and other sexually reproducing creatures with a much needed biological flexibility in a world of unstable environments and countless threats to survival.

Although sexual reproduction creates the diversity that enables us to be flexible survivors in a constantly changing world, this adaptive flexibility comes with an exceedingly high price. Genetic transport vehicles like you and me, the proud carriers of the immortal genetic code book of life, are disposable. But why should this be so? It seems like a terrible waste to invest so many resources into building and maintaining a body that is only going to be abandoned early in life. To answer this why question, we must discard our soma-based biases and view the world from the perspective of our immortal genes.

To do this, we are going to swap places with our genes. First of all, think of your body as a leased car. You, as the genes, are both the creator and the driver of this genetic transport vehicle. Under the terms of the lease, you are responsible for maintaining the car. At first everything is fine; the monthly cost of the lease is fixed, and the maintenance costs are low or nonexistent. As time goes by, however, the maintenance costs begin to rise as the vehicle begins its inevitable accumulation of damage caused by wear and tear. Your goal, as genes, is to make it to the next generation. So you decide to trade the vehicle in for a new model before incurring the inevitably high expenses of maintaining the old model. Do not think that this process requires conscious thought or that other sexually reproducing species have not already adopted the same strategy. Your genetic competitors that adopted this approach are still around; those that did not are long gone.

For sexually reproducing organisms, the river of DNA that flows through time runs dry when no offspring are produced within any generation. Reproductive failure is not like an athletic event where there is another race tomorrow or a next time at bat; it means the final death of a genetic lineage that cannot be undone. Diverting resources away from producing new transport vehicles in order to create an immortal soma is genetic suicide. The risk of death for our ancestors was exceedingly high very early in life. In such a hostile world, unrepaired damage to the soma cannot be avoided. Even if attainable, the resource costs required to create and maintain a perfect soma would be prohibitive. In the final analysis, you and I are expendable. Death is the price we pay for the immortality of our genes.

Accepting the inevitability of death still does not answer the one burning question we all think about. When will I die? Maybe this question will be answered with authority once scientists can not only read but also understand the genetic code book. All that can be said today is that this question remains an unsolved problem with an answer that is currently nowhere in sight. Paradoxically, if we consider human *populations* rather than specific individuals, we know a lot about the when question. For example, why isn't the average human lifespan 5 years, or 500 years, instead of the 78 years we now observe in developed nations? If our perspective is broadened even further to include other species, we can begin to shed a brighter light on the question of why we die when we do.

Why we die when we do, both as individuals and populations, is ultimately determined by answers to the question, why do we age? A handful of brilliant scientists have provided insightful answers to this most important of all biological questions.

EVOLUTION THEORIES *of* AGING

In 1891, the famous German biologist August Weismann formulated one of the earliest evolutionary theories of aging. His goal was to explain why different kinds of animals live as long as they do. Weismann used a simple graph to illustrate how longevity increases as body size increases—large species like elephants live much longer than small species like mice. He reasoned that large animals acquire the genetic ability to live longer because it takes more time to build a large body. Arguing that large animals live longer *because* they are larger is an example of circular reasoning. Faulty logic, however, was not why Weismann eventually rejected this explanation for longevity. Instead, he concluded that body size alone could not provide a general explanation for longevity because there were just too many examples of small fish, reptiles, birds, and mammals that lived far longer than larger animals.

Weismann also wondered whether "rate of physical activity" (what we call metabolic rate today) might explain longevity. Once again, there were species that simply did not fit the rule. However, in trying to link rate of activity to longevity, Weismann revealed a concept that would play a central role in future explanations for why species lifespans differ so dramatically. Weismann believed that the purpose of individuals is to perpetuate the species. Weismann's valuable insight was that "the duration of life is forced upon the organism by causes outside itself, just as the spring is fixed in its place by forces outside the machine, and not only fixed in its place, but chosen of a certain length so that it will run down after a certain time." In other words, he was telling us that the biology of organisms that determines longevity is shaped by the environment within which they live.

The beauty of this concept is that it is counterintuitive. Most of us think of longevity as being determined by a combination of genetic factors and choices made about lifestyle. Weismann's explanation for longevity was far different. He suggested that species's lifespans are determined by environmental forces that seem at first glance to be totally unrelated to the biological and behavioral forces most of us associate with aging. Among his environmental forces were such factors as infectious diseases and the small injuries that accumulate through the course of life.

In Weismann's world, life is a double-edged sword. Reproduction generates the new bodies needed to survive in a hostile environment, and death is the way nature discards the old, worn-out bodies. According to Weismann, organisms are highly adapted to their environments, with nature favoring those characteristics of an individual that serve the primary purpose of life—reproduction. If the lifespan of a species becomes too short in a particular environment, natural selection will favor its extension. Conversely, excess longevity will disappear in hostile environments. Just over a hundred years after Weismann, the idea that individuals are disposable once their reproductive role has been accomplished remains a cornerstone of modern theories on the evolution of aging.

In yet another forerunner of the modern view, Weismann described a cellular mechanism that could explain how organisms "run down after a certain time." Death, according to Weismann, takes place "not in the waste of single cells, but in the limitation of their powers of reproduction. Death takes place because a worn-out tissue cannot for ever renew itself, and because a capacity for increase by means of cell-division is not everlasting, but finite." Weismann's notion was not confirmed until the biologist Leonard Hayflick, seventy years later, demonstrated that somatic cells can only replicate a limited number of times. This phenomenon has been linked recently to a progressive shortening of a region of DNA at the ends of chromosomes known as the *telomere*.

Weismann suggested that reproduction was necessary to perpetuate the species, while aging, decay, and death were the means by which nature eliminates the old who compete with the young for limited food resources. This was the accepted position on the roles of reproduction and aging from the time of Darwin until the middle of the twentieth century, when biologists began to declare that the individual, not the species, is the entity upon which evolution acts. Today, with our much expanded knowledge of genetics, it is recognized that the gene is the basic unit of evolution. In fact, contrary to what frequently appears in the popular literature, modern theory holds that evolution is totally indifferent to whether species survive—a scientific position supported by a fossil record dominated by extinction.

Jumping ahead to 1952, the British Nobel laureate Sir Peter Medawar published a paper that, like Weismann before him, stressed the importance of deaths occurring early in life from causes unrelated to aging. Medawar's great insight was a recognition that prior to about the middle of the eighteenth century, very few people lived long enough to experience the consequences of aging because most died at earlier ages from infectious and parasitic diseases. He used the concept of natural selection to provide one of the first modern explanations for why aging exists.

Sexual reproduction causes the genetic code book to be reshuffled in each new generation. Each of us is unique because we all carry slightly different forms of the same genes (*alleles*)—variation in eye color is a good example. Medawar used variation among individuals in the age at which alleles are expressed to take us one step closer to understanding how and why aging occurs. Consider two hypothetical alleles, both of which cause a fatal disease but at different ages. One allele causes death at age twenty, while the other allele waits until age twenty-five before it kills. Which allele will appear with greater frequency in the next generation? The answer is the allele expressed at age twenty-five. Those individuals carrying the other allele would, on average, die at younger ages, and therefore have fewer children carrying that allele into the next generation. In each succeeding generation, the frequency of individuals carrying the later expressing allele should increase relative to those with the early expressing form.

Medawar saw natural selection acting in this way on all the genes in our body. The genes that are more harmful to survival and reproduction are figuratively "pushed" by natural selection to later and later ages, where they have less of an effect on reproduction. Eventually, over many generations, harmful alleles will tend to have their age of expression accumulate at or near the end of the reproductive period. Natural selection cannot simply eliminate these alleles because they have already been passed on to the next generation. In a moment of figurative zeal, Medawar called the post-reproductive period of life a "genetic dustbin"—a time in the lifespan where many of the damaging and lethal genes can be expressed without penalizing the reproductive productivity of the organisms carrying them. For example, if you carry a gene that causes lung cancer but the gene is not expressed until you reach age 120, you probably would not be too concerned, knowing that your chances of living that long are slim. Natural selection would not be too concerned either.

Medawar's insight was to divide the lifespan of a species into stages. In the human case, the stages are the pre-reproductive period (between birth and puberty, about age twelve), the reproductive period (mostly between ages twelve and fifty), and the post-reproductive period. Natural selection is highly efficient at reducing the frequency of lethal alleles that are expressed in the pre-reproductive period because these genes kill before the story of life has a chance to be passed on. However, once puberty is reached, the ability of nature to remove dangerous alleles declines rapidly because this is the moment when reproduction begins. Genes, whether good or bad, are beyond the reach of natural selection when they are expressed at ages beyond the reproductive period. By coupling the time of gene expression during the lifespan to the action of natural selection, Medawar provided the scientific world with a genetic explanation for why aging exists and how aging works.

Medawar also recognized the implications of his theory for the real world. He realized that advances in medicine and public health were permitting unprecedented numbers of people to survive into the post-reproductive period for the first time in human history. Extended survival means that the diseases in the genetic dustbin have the opportunity to be seen or expressed in a greater number of people. Huntington's chorea and Lou Gehrig's disease are classic examples of the kinds of diseases to which Medawar was referring. These incurable neurological diseases are inevitably passed from generation to generation because their symptoms do not appear until the ages of thirty-five to forty-five. Medawarian diseases, as we refer to them, are not some cruel hoax perpetrated by Mother Nature. They are the predictable price we must pay for human mastery over the infectious and parasitic diseases that in the past have denied access to old age for the majority of the population. In the poetic words of Medawar, aging can be revealed "only by the most unnatural experiment of prolonging an animal's life by sheltering it from the hazards of its ordinary existence."

Another piece of the puzzle explaining why aging occurs was provided in 1957 by George Williams, one of the most influential evolutionary biologists since Charles Darwin. Williams asked a question similar to that of Medawar: how is it possible for genes with adverse health effects to increase in a population? He answered this question by developing an innovative twist to an established genetic phenomenon. Geneticists had long known that an individual gene can perform more than one function at any moment in time. There is not a single gene whose sole function is to create an eye and another individual gene that makes a liver. If this were the case, then considerably more than 80,000–100,000 active genes would be required to build and operate a human body. This multiple behavior of genes is so well established among geneticists that it has its own special name—*pleiotropy*.

Unlike the "harmful" genes described by Medawar, Williams envisioned perfectly normal genes that were involved in critical growth and development processes early in life. His innovation was to suggest that genes acting normally early in life could act abnormally later in life. In other words, Williams argued that genes function *pleiotropically in time*. Nature would favor the accumulation of genes that do beneficial things early in life even though they might do harmful things late in life because, under normal conditions, most animals do not live long enough for the harmful effects to cause a problem. Williams created the descriptive term *antagonistic pleiotropy* to refer to his concept of genes that do beneficial things early and harmful things later.

Like the predictions of Medawar, Williams's antagonistic pleiotropy hypothesis is consistent with recent findings on the behavior of aging-related diseases. Molecular biologists are finding that many of the genes responsible for cancer late in life are intimately involved in the regulation of cell growth and differentiation early in life. The genes that have such devastating effects late in life when expressed as diseases like Alzheimer's appear to be unaltered in form from when they serve a useful function early in life. Antagonistic pleiotropy is one reason why it will be nearly impossible to genetically alter human beings for the sole purpose of achieving greater longevity. We might be trading off adverse health consequences early in life for a few extra years late in life—assuming we even survived to older ages. As Medawar said, the damaging effects of genes that cause Alzheimer's disease, and many of these diseases and disorders that all of us recognize as consequences of aging, can only be observed under the unnatural condition of surviving to old age. The fact that the mechanisms for these dreadful diseases have yet to be worked out make the insights of Williams even more remarkable.

Finally, we turn to an important concept in the evolution of aging developed in the late 1970s by the British biologist Thomas Kirkwood and his colleagues. Like those before him, Kirkwood argued that environmental sources of mortality play an important role in determining when aging occurs. These environmental (external) causes of death determine the age window within which organisms are likely to survive and reproduce. The reproductive window, in turn, determines how effective natural selection can be in weeding out harmful alleles from a population. In an evolutionary sense, Kirkwood argued, an organism must "choose" between an investment in producing offspring and an investment in protecting and maintaining the soma. In the economic view of life developed by Kirkwood, aging and the disposal of the soma (death) are inadvertent consequences of investments in reproduction made necessary by a hostile environment. Kirkwood reasoned that it would be foolish to invest precious biological resources toward achieving immortality for the soma in a hostile world where immortality is not possible anyway.

These evolutionary theories of aging have profound implications for the current state of human existence. Everyone alive today has inherited a genetic legacy from a long line of ancestors who lived under far more hostile environmental conditions than we experience today. The conse-

quences of this genetic legacy include the capacity to reproduce by twelve years of age and the fact that most reproduction
is over well before age forty. The evolution theories of aging
from Weismann to Kirkwood suggest that natural selection
has shaped human biology such that aging and death become
increasingly likely by the time we reach our forties. In truth,
aging begins the moment the engine of life is switched on at
fertilization but does not become noticeable until the fifth
decade of life, when the risk of death is so high that it begins
to strike close to home.

Death Genes *and* Race Cars

There is one more misconception that still flares up occasion-
ally in the popular press. Once nature abandons us for a new
leased vehicle (our children), why doesn't a death gene take
our life quickly and efficiently in the middle of the night, after
healthy life is no longer possible? Why isn't nature kind and
gentle enough to eliminate the pain and suffering that often
accompanies old age and the process of dying? Why doesn't
Mother Nature take a more active role in eliminating the old?

There is an easy way to dispense with this issue—one that
is a lot simpler than discussing genes and alleles. Consider a
race car. In order to build the right kind of car, you need to
know something about the race itself. How long is it, what are
the road conditions (cement, asphalt, gravel, or dirt), and will
you encounter extreme environmental conditions like high
temperatures, rain, or snow? For long races, a durable car is
needed, one that can withstand more wear and tear than those
designed for shorter distances.

Your job is to build a car for the Indianapolis 500. Since
you and all of the other engineers already know about the con-

ditions of the race, you adopt similar, but not identical, strategies for constructing your cars. Durable parts are going to be
needed so that your car can survive the harsh conditions of this
long and grueling race. If your car fails in the test runs, you
bring it into the shop, find the faulty part, and reengineer it to
last longer. One major problem you confront is that it is not
possible to predict exactly what might go wrong during any
given race. Your solution is to engineer the critical parts of the
car to be more durable in order to ensure that they last beyond
the end of the race. This is equivalent to building a bridge to
withstand more weight than would ever be encountered under
ordinary conditions. However, because your car is not normally operated much beyond the end of the race, you are not
concerned with things that might go wrong after 500 miles.

Now that the car is built, you are also going to be the
driver. Because you are behind the wheel, you have the power
to exercise some control over the car during the course of the
race. The objective, of course, is to win the race. The design of
the car is determined before you sit in the driver's seat, but
how the car is driven is ultimately your choice.

Now let's conduct an experiment. Instead of turning the
engines off at the end of the race, everyone is going to continue racing around the track until every car has broken down.
For the first time, we will have the opportunity to inspect parts
that cause the cars to stop running because of accumulated
wear and tear. Naturally, differences in original design and
how the cars are driven will cause them to vary in how far they
travel before breaking down. Some cars fail shortly after 500
miles, a few well-designed cars operate far beyond the normal
end of the race, and the majority inevitably fail somewhere
between these extremes.

As it turns out, the failure times of mechanical devices like
cars, bicycles, washing machines, and televisions follow a pre-

dictable pattern—similar to that of living organisms. The manufacturers of cars and other mechanical devices are acutely aware of these failure patterns, which lead them to create short and low-risk (for them) warranty periods. Assuming competent engineering, most of the race cars are probably comparable in their ability to complete the race. However, under the new condition of extended operation, the failure times of the cars will differ. Variation in failure times reflects variation in engineering and the random or chance locations of damage that accumulated during the extended race. Upon inspection, it will be found that a common set of "weak links" emerges for the cars used in our unusual race. The reason for this commonality is that the conditions of the Indianapolis 500 force the engineers to use a limited number of car designs if they intend to compete successfully in this race.

There are several important points to be drawn from the car analogy. First, the car begins accumulating damage as soon as the engine starts, and the damage occurs no matter how carefully the car is driven. Second, although there is an element of chance involved, the locations of the damage that cause failure tend to be concentrated within a few specific parts (weak links). Third, the driver does have control over how the car is operated. Fourth, there is no incentive to design a car that lasts much longer than the prescribed distance of the race. The costs of designing and constructing an immortal race car would be so enormous that there would not be enough projected profit to induce sponsors to build the car in the first place. Finally, the most important lesson to remember is that the race cars were not purposely engineered to fail. They were simply not designed for extended operation.

Now let's repeat this thought experiment using sexually reproducing organisms. Natural selection is the engineer and the end of the race is measured in time, not distance. From an evolutionary perspective, the race is to reproduction. In the realm of evolutionary theory, reproduction is defined not just as the age range within the lifespan during which reproduction occurs, but for some species it can include a period of time for child rearing and even grandparenting. The reason that nature has not forgotten those of us who are beyond the age of generating offspring is that we can still help get more copies of our genes into future generations by enhancing the reproductive success of our children and grandchildren.

For a race car, the capacity to finish the race is critical. For sexually reproducing species, transferring the genetic code book to the next generation is the goal. This goal cannot be achieved without survival to sexual maturity (puberty). The age of sexual maturity for any species requires passage through a highly predictable series of developmental stages whose timing follows the rhythm of biological clocks that are under considerable genetic control. As evolutionary biologists have suggested, the tempos of these clocks are set by natural selection operating through a hostile environment that rigorously defines an age window when successful reproduction can occur. Going back to our analogy, this is equivalent to an engineer using knowledge about the conditions of a race in order to design and build a car appropriate for that race.

The IMPLICATIONS *of* EXTENDED SURVIVAL

Everything that is known about life on earth indicates that under normal conditions, the vast majority of organisms die early in life. Somehow your ancestors survived this gauntlet of death and were able to pass the story of life to the next generation. It is important to realize that we are talking about surviving long enough to reproduce, not grow old. For any species, including humans, only a very small fraction of the individuals ever born have lived long enough to experience the effects of aging.

Now assume that a way has been found to permit most individuals of our hypothetical population to survive beyond the age range of the typical reproductive period. In light of the race car analogy, what will happen to the individuals who are surviving to unprecedented older ages? For the first time, we get the dubious honor of observing failures in an array of essential components of the body that would not be observed under normal conditions because survival to old age is rare.

For humans, our pets, and laboratory and zoo animals, our hypothetical experiment has become a reality. The common diseases of aging now observed in humans and these other animals are a direct result of survival into the postreproductive period of the lifespan occurring with a frequency that is unprecedented in the history of life on earth. Aging and the diseases of aging appear at such predictable points within the lifespan of individuals that they seem to be following a tempo of life that is being orchestrated by biological clocks.

Biological Clocks

Biological clocks conjure up images of windup toy soldiers and model airplanes powered by rubber bands. Wind these mechanical devices up, and the amount of time that they stay in motion—their duration of activity—is determined by the size of their springs or rubber bands and how tightly they are wound. Similar principles apply to the biological clocks that influence longevity. The Taoist concept of "vital breath" echoed in the philosophies of Bacon and Cornaro captures this notion of something lost or used up during the course of life.

The image of biological clocks winding down would seem to conflict with our earlier argument that organisms do not have a fixed limit for the duration of life. After all, a mechanical clock stops ticking after a fixed amount of time has transpired. There is a solution to this apparent paradox. The metronomes of life are biological processes that are governed by genes, not by metal springs, gears, or weights. Natural selection can only affect the longevity of individuals indirectly by exerting a direct influence on genes whose functions have an impact on reproductive success—for example, biological attributes that contribute to health and vigor early in life.

Death genes, if they existed, would behave in a radically different way. These genes would have to switch on at some predetermined post-reproductive age in order to terminate life. The requirement that death genes become activated at ages beyond the reproductive years means that evolution could not give rise to them. Because natural selection can only influence genes that affect the production of offspring, evolution could not have led to either death genes or the clocks needed to turn them on at specified ages within the post-reproductive period of the lifespan. The inescapable conclusion is that the biological clocks that are present within organisms exist for one reason only—to support life, not destroy it.

By definition, reproductive success requires that survival extend to the ages when reproduction is possible. This implies that biological clocks play a critical role in the numerous processes that transform a fertilized egg into an "adult" capable of passing on the genetic story of life to the next generation. They synchronize the developmental processes that construct a fetus in the womb, coordinate the physical and mental changes that occur as an infant becomes a young adult, influence the timing of puberty, control when eggs are released from the

ovary, and determine when the uterus is ready to nurture a fertilized egg. Although considerable genetic control is implied by
the regularity and predictability of their beat, biological clocks
can also be modified by environmental forces. Delayed puberty
in females caused by malnutrition is a well-documented
example of such an environmental effect. Poor environments
forced upon a developing fetus by its mother during gestation
can delay developmental processes in childhood, and may even
influence the risk of experiencing degenerative diseases later
in life. The nearly universal regularity and predictability of the
stages of growth, development, maturation, and reproduction
points to the profound influence of biological clocks on the
tempo of life. How is the tempo of the biological clocks in people determined? Why do puberty, menopause, disease, and
death occur when they do? The answers to these questions lie
in our distant past.

HUMAN LIFE *at* ITS BEGINNING

It is tempting to be nostalgic about the past. Images come to
mind of a time when the air was fresh, water was pure, man-
made environmental toxins were nonexistent, game was plentiful, and the trees were laden with fruit. This image is far from
the reality. In fact, the environments that shaped mankind
were decidedly dangerous and hostile. By examining these
environmental conditions, we can begin to understand why
puberty, menopause, disease, and death occur when they do.

Nobody knows for sure what daily living conditions were
like 130,000 years ago when modern humans arose, but the
archeological record provides some clues. Living in one place
was not an option for our early ancestors because agriculture
had not yet been developed. Their lives were largely devoted

to just one thing: staying alive. Most of their time would have been spent searching for food, avoiding predators, and finding shelter from the world in which they lived. The frequency of bone and joint deformities in fossilized remains suggests that bone fractures were common and often healed improperly. Indications of crippling joint diseases in the jaws, spinal columns, hips, and feet are also common. Extensive tooth loss and holes in the jaw signal the presence of abscesses and infections of the bone. When enough bones are found to permit an estimate of chronological age, individuals older than thirty to forty are rare. The forensic evidence is clear: life was extremely hazardous at the dawn of our species.

The historian William McNeill points out in his book *Plagues and Peoples* that our ancestors shared their bodies and their homes with an impressive array of disease-causing microbes—fungi, protozoa, bacteria, and hundreds of viruses spread by rodents, mosquitoes, mites and other insects. Our ancestors also lived in close proximity to large predators such as lions, tigers, and crocodiles—and there can be little doubt that children and the infirm were often on the menu of these large predators.

Common ailments like broken bones and teeth, strained muscles, cuts, animal and insect bites, puncture wounds, and infections would have been life-threatening events. It is probable that as many as half of all babies born died before reaching their first birthday, and the mothers who delivered them did so without any kind of medical intervention—often dying in the process. Of the children who did survive, only a small percentage would have lived long enough to have children of their own.

There is no doubt that our earliest ancestors were severely challenged by the world they lived in. In the final analysis, there can be no better indicator of the severity of these challenges than the average age at which they died. Of those who survived childhood, few lived past their twentieth birthday and a forty-year-old would have been considered an elder member of society.

LIFE *in the* NINETEENTH CENTURY

Moving ahead to the late nineteenth century and to large metropolitan areas, our great-grandparents' and grandparents' daily activities were probably quite similar to those of city dwellers today. However, lurking under the surface of this similarity was a world far different from our own. One hundred years ago, the air was visibly polluted and the living and working conditions were conspicuously harsh.

An example of just how harsh life was then is the painful reality that up to one-third of the babies born in a large city at that time would have died from an infectious disease. The obituaries were also filled with sad accounts of young women who died during childbirth. Large, overcrowded populations created an almost ideal environment for sustaining a variety of infectious disease organisms. The diseases that these organisms caused are known to us primarily by their reputation—cholera, diphtheria, influenza, malaria, polio, smallpox, tetanus, and tuberculosis, to name a few. Nearly everyone who lived at that time, rich or poor, was likely to have experienced a personal tragedy linked to one or more infectious diseases. Some of these diseases still take a heavy death toll in parts of the developing world.

Refrigeration as we know it did not exist. Foods kept in faulty ice boxes turned rancid within hours. The stench of human feces, urine, and horse manure filled the streets because public sanitation, waste disposal, and automobiles had not yet become a routine part of life. Public water supplies, including those derived from streams and wells, were often contaminated by parasites, the bacteria from human waste, and other waterborne diseases like the deadly cholera. As a result, chronic cases of intestinal parasites and diarrhea were common. Malnutrition was commonplace because knowledge of nutrition was limited— especially the nutritional needs of infants, children, and women who were either pregnant or nursing.

Central heating and air conditioning did not exist and many homes lacked proper insulation. In the United States, whether people lived in northern climates like New York State or southern climates like Texas, homes provided little more than partial protection from climatic extremes. Occupational safety standards did not exist. As a result, working environments were often hazardous—especially in factories, service industries, and agriculture. Sanitary conditions in hospitals were primitive by today's standards. This meant that hospitals were often places where infectious diseases were easily spread and routine health concerns like broken bones and appendicitis often turned into lethal conditions.

Living conditions such as those just described are something that many of us in developed countries today just read about in history books. Unfortunately, we know that the devastating effects they have on human lives are not exaggerated because conditions like these still exist in many parts of the world. It is undeniable that our grandparents, great-grandparents, and their ancestors lived in a world far more hazardous than the one we experience. As such, nostalgic images of pristine rivers and lakes, pollution-free air and simpler lives, are appealing fantasies. The reality is that as recently as just one hundred years ago, the environment was harsh, living and working conditions were dreadful, and life expectancy was low (between 45–49 years) by comparison to today.

The IMPLICATIONS of a GENETIC LEGACY

From 130,000 years ago to 100 years ago, the environment did not permit the average person to live much beyond the age of forty-five. Although the specific threats to human life may have changed over the millennia, the effect on survival has been the same: most people died young. These were the environmental conditions in which natural selection forged the genome of our ancestors—genes that have been passed down through the ages to everyone now living. The biological responses of our species to those conditions are inscribed in our own genetic code books. Everyone alive today is the most recent link in an unbroken chain of survivors that extends back not only to the origins of modern humans but to the origins of life on earth. As such, each and every one of us carries a legacy of genetic responses to past environments—environments that posed far greater challenges to survival than those experienced by most people today.

In a hostile environment, one of the most crucial responses that an organism can make is the timing of reproduction. This age is defined by when sexual maturity occurs. Delaying the age of sexual maturity increases the risk that death will intervene before reproduction can occur. High mortality risks create evolutionary pressure to attain sexual maturity as early in the lifespan as is physiologically possible. Although common patterns of reproduction can be identified, every species has evolved its own unique reproductive strategy to cope with a challenging environment. These strategies, of course, are not conscious decisions but are, instead, biological responses dictated by natural selection and the genetic history of the species. The American horticulturist Luther Burbank describes this history as a genetic legacy, one that portrays genes as a form of stored environment. For example, field mice produce litters that are capable of reproducing within thirty days of their own birth—a necessary response when so many organisms consider a mouse a meal. Conversely, the protective shells of tortoises and turtles permit these creatures to delay maturity and first mating for decades in some cases. The human genetic legacy gives rise to sexual maturity occurring somewhere between ten and sixteen years of age. Although environmental factors such as the availability and quality of food are known to influence this timing, genes play a central role in the transformation of a fertilized egg into a reproducing adult.

The transformation of a fertilized egg into a sexually mature adult is a complicated process of growth and development. The age of sexual maturity establishes what may be thought of as a production deadline for completing the biological response of a species to the challenges posed by the environment within which it exists. All of these processes are orchestrated by biological clocks that set the tempo for genetic programs of growth and development—integrated programs of amazing complexity, all of them a legacy of genetic responses to past environments.

In order to reproduce, an organism must survive to reproductive ages. In hostile environments, species have developed a wide range of behavioral tactics, such as decisions to either fight or flee, that improve their chances of survival. In addition, organisms also possess a remarkable ability to repair damage that accumulates during the course of everyday life. There have been many debates over the contribution that genes make to behavior, particularly human behavior, but there is no debate among scientists over the central role that genes play in maintaining the biological integrity of the organism that carries them.

From ANCIENT PERILS *to* MODERN DISEASES

Why do most people today die from diseases that are different from those that killed most of our ancestors? Why do these "modern" diseases kill people when they do? What is the relationship between the diseases that kill people today and the ancient perils that threatened our ancestors?

Biological clocks are systems of genes that perform and regulate the processes of growth and development designed to culminate in sexual maturity and reproduction. Because these processes and the genes that regulate them are designed for maintaining and nurturing life, evolution cannot design genes for the purpose of killing people—neither the diseases that killed our ancestors, nor those that kill people today. In other words, genes that cause sickness, disease, and death are an inadvertent by-product of evolutionary neglect, not the end product of evolutionary intent.

The genetic code book carried within the human body today is in every important way the same one that was carried by our ancient ancestors. This genetic legacy of biological responses to the hostile environments of our ancestors is directly responsible for the diseases we experience today, and the ages at which they appear. The biological clocks that we have inherited from the past build bodies that can reproduce by around twelve years after birth, and begin revealing the consequences of extended survival around forty years after birth.

Because genes are designed to ensure that animals survive long enough to reproduce, once sexual maturity has been achieved, the genetic programs designed to sustain life begin to operate less efficiently. Today, most people survive beyond the ages when when most reproduction typically occurs. Significantly, this extended survival clearly demonstrates that

from an evolutionary perspective, human beings are overengineered—surviving longer than is necessary to produce the offspring that carry their genes. There is no need to improve upon
a body design that is already overengineered. This is why life-
sustaining systems and biological clocks wind down, and ultimately it explains why people today die when they do.
Surviving into the post-reproductive period permits the diseases and disorders of human aging to be observed.

Why does living past the reproductive period lead to the
specific diseases we experience today? Just as the weak links
in a race car are revealed by operating it beyond the end of the
race, the weak links of the human body are exposed when people survive beyond the reproductive period. In humans and
other mammals, the weak links include a cardiovascular system that clogs up and is subject to wear and tear (heart disease
and stroke), a weakening immune system (infectious diseases), DNA that accumulate damage (cancer), a skeletal system that becomes more brittle at later ages (osteoporosis),
and deteriorating sensory systems (vision and hearing loss).
Our early ancestors were subject to the same processes of
degradation, but most of them did not live long enough to
experience them.

Despite all the remarkable gains in longevity that have
been made through human intervention, there remains variation in the ages at which disease and disability are expressed.
There are several reasons why the ages at which sickness and
death occur vary from person to person. Genetic uniqueness
is an inevitable by-product of sexual reproduction. Age variations among individuals also arise because there is a random
component to when, where, and how much damage our bodies
accumulate. Finally, variation in lifestyle choices among individuals contributes to the timing of disease expression. No
matter how sophisticated human technology becomes, in

every new generation some children are destined to die before reaching puberty while others may live only a short time longer. A few will be born with the right combination of good genes and good luck to survive to extreme old age—some even retaining their health along the way. As for the vast majority of us, for better or for worse, we will experience health, disease, and eventual death somewhere between these two extremes.

THE PREVAILING VIEW among many advocates of extreme prolongevity is that aging is just another disease and lifespans have grown shorter over time because lifestyles have become progressively more decadent. Everyone is born with the potential to live a long life, and the diseases we experience are a product of poor lifestyles, a harmful environment, and the depletion of one or more "vital" substances. Some of these advocates claim that they can supply these lost or depleted substances and restore youth, most often at a substantial financial cost. Return to the healthful ways of our early ancestors, or ingest their *elixir vitae*, they say, and we will all be graced with the gift of great longevity. This chapter should make it clear that this notion of prolongevity is based upon a combination of folklore, pseudoscience, misconceptions, and false reasoning about how and why people age.

The process of aging and the diseases that accompany it are not an exclusive by-product of decadent lifestyles, as many would have you believe. Instead, they are in large measure a product of a genetic legacy inherited from our ancestors, and the expression of that genetic legacy by people who are now living well beyond the warranty period for human bodies.

Human biology has been forged by environments that

have been hostile to life for nearly the entire history of our species. In the last one hundred years, we have learned how to make these environments far less hazardous. The hardy constitution that was needed in the past just to survive long enough to reproduce is what allows people today to live far beyond the ages that our biology defines as the reproductive years. A forty-year-old man would have been an elder in the society of our early ancestors. Today, a forty-year-old man is considered to be in the prime of life. A sixty-year-old woman was an oddity a few thousand years ago. Today, the death of a sixty-year-old woman is considered premature and centenarians have become commonplace. It is not *Homo sapiens* or the process by which humans age that has changed over time; it is the perception and meaning of old age that has changed so dramatically.

It is appealing to be nostalgic about the good old days before automobile accidents, plane crashes, radioactive contamination, pesticides, air pollution, toxic and chemical wastes, handguns, and drugs. Instead of an Antediluvian view that looks to the past for a utopian world free from aging and disease, everyone should realize that in terms of health and longevity, the world in which we live today is the best that has ever existed.

CHAPTER 3

—✕—

Life Expectancy

It's not that I'm afraid to die,
I just don't want to be there when it happens.

WOODY ALLEN

NOBODY KNOWS HOW LONG any single individual will live. It is possible, however, to make an educated guess by calculating a statistic that demographers and actuaries call life expectancy. One approach to estimating life expectancy is to calculate the average age at death for a large group of people who have been monitored from their birth until everyone in the group has died. Calculating human life expectancy in this way is not feasible because it would require waiting more than a century before the calculation could be performed. In fact, the scientist trying to make this calculation would almost certainly die before obtaining the answer.

Fortunately, there is a simpler but somewhat less precise way to estimate life expectancy. For years, government agencies in the United States and other countries have kept track of the number of births and deaths that occur each year, as well as how old each person was when he or she died. This information, coupled with Census data on how many people are alive at each age, permits scientists to calculate both the risk of dying and the probability of surviving another year for people of every age. These survival probabilities can then be used to create what is known as a life table. The life table contains estimates of life expectancy for people at any age. These estimates are called "period life expectancy" because they are based on survival probabilities that were observed at a specific moment in time, like the year 2000. Period life expectancy at birth is what you typically see reported in the news, and it is what extreme prolongevists and some scientists are referring to when they predict that life expectancy will rise to 100 years or higher in the twenty-first century.

The life table for the United States provided below (see the appendix for a more complete table) contains three sets of numbers: estimates of life expectancy at every age for males and females presented in both years and days, and the probability of surviving to your next birthday. Although the table was calculated for the population of the United States in 2000, the values in the table are reasonable estimates of life expectancy for people currently living in any developed nation (a few days could be added to the tabulated values in order to adjust for the small gains in life expectancy that have occurred since 2000).

Notice that life expectancy can be calculated at any age. For male and female babies born in 2000, the expected remaining years of life were 73.5 years and 79.6 years, respectively—in other words, between 26,000 and 29,000 days. If you are currently 50 years old, the table indicates that you are expected to have 10,044 days of life remaining if you are male, and 11,651 days of life remaining if you are female. Because these are averages, the tabled value for 50-year-old males, for example, also means that half of them are expected to live longer than 10,044 days, and the other half are expected to die before this many days has passed.

It is important to remember that a life expectancy is an average for a population. One consequence of this is that no matter when a person dies, a life table will always indicate that the individual died with expected days of life yet to be lived—referred to as "residual life expectancy," or remaining days of life. So, when Frank Sinatra died in 1998 at the age of 82, according to the life table, he had 6.5 years or 2,374 days of life remaining. Similarly, when George Burns died just short of his 100th birthday in 1996, he had a residual life expectancy of about 2.3 years or 804 days. Residual life expectancy has been misinterpreted by some people to mean that the death of an individual is always premature—a fallacy that arises from using population averages to draw conclusions about specific individuals.

If the second column of the life table alarms you, you might look over at the third: the probability of surviving to your next birthday. Notice that during the first seven decades of life, these numbers exceed 99 percent for both sexes, and they do not drop below 80 percent until after 90 years of age.

The apparent paradox between a depressingly few days of life remaining, and great odds that you will see your next birthday, is resolved by understanding another aspect of life tables. In actuarial jargon, the numbers that you see in the table are referred to as "conditional." This simply means that they are influenced by how long you have already lived. To illustrate, a girl born in 2000 is expected to live 79.6 years or 29,074 days, but a woman who has already lived 80 years or 29,220 days is expected to live another 3,579 days beyond the estimate for a newborn. Why does this happen? Once again, the answer lies within the mathematical nature of averages.

Current Age in Years	MALES			FEMALES		
	Years of Life Remaining	Days of Life Remaining	Probability of Living to Your Next Birthday	Years of Life Remaining	Days of Life Remaining	Probability of Living to Your Next Birthday
0	73.5	26,846	99.3	79.6	29,074	99.4
5	69.1	24,239	99.9	75.2	27,467	99.9
10	64.2	23,449	99.9	70.3	25,677	99.9
15	59.3	21,659	99.9	65.3	23,851	99.9
20	54.6	19,943	99.9	60.4	22,061	99.9
25	50.0	18,263	99.8	55.6	20,308	99.9
30	45.4	16,582	99.8	50.7	18,518	99.9
35	40.8	14,902	99.8	45.9	16,765	99.9
40	36.3	13,259	99.7	41.2	15,048	99.9
45	31.9	11,651	99.6	36.5	13,332	99.8
50	27.5	10,044	99.5	31.9	11,651	99.7
55	23.3	8,510	99.3	27.4	10,008	99.5
60	19.4	7,086	98.7	23.2	8,474	99.2
65	15.8	5,771	97.8	19.3	7,049	98.7
70	12.7	4,639	96.8	15.7	5,734	98.0
75	9.8	3,579	95.2	12.3	4,493	97.0
80	7.3	2,666	92.4	9.4	3,433	95.2
85	5.4	1,972	87.9	6.8	2,484	92.0
90	3.9	1,424	81.6	4.8	1,753	86.5
95	2.9	1,059	73.7	3.4	1,242	78.4
100	2.2	804	65.6	2.5	913	69.8
105	1.6	584	56.1	1.8	657	59.6
110	0.5	183	45.0	0.5	183	45.0

Like kids adjusting their relative position on a teeter-totter in order to make it stay level with the ground, an average is the balance point of the collection of numbers it represents. As such, life expectancy at birth is lowered by those who die at younger ages. As death continuously removes people from a population, the remaining survivors become progressively dominated by those with increasingly more favorable longevity potentials—people whose life expectancy at birth was greater than the average for all babies born in that year. Although the statistics of a life table must be interpreted with caution, they reflect the biological reality that a human population is an assemblage of genetically unique individuals with a wide range of longevity potentials that are eventually revealed by survival and death.

Period life expectancy is an imperfect measure because it fails to account for the improved survival brought about by advances in the biomedical sciences. A brief example will clarify this point. Assume that we want to know the probability that a male baby born in the year 2000 will survive to his 50th birthday. To calculate this probability, we use the survival rates in the year 2000 for males at ages between birth and age 50. These survival probabilities are then applied to male babies born in the year 2000 for each of the first 50 years of their life—years that have not yet been lived by the newborns. Experts believe that these probabilities are too low because advances in medicine that improved survival in the past are expected to continue in the future. As a result, males who reach age 50 in the year 2050 should have a greater chance of reaching their 51st birthday than their counterparts did who were 50 years old in the year 2000. When death rates are declining as they are now, most people will live slightly longer than the calculated period life expectancy. Nevertheless, throughout this book and in popular usage, the term *life expectancy* almost always refers to the calculation of period life expectancy at birth.

The METHUSELAH TRAP

As a child growing up in Detroit, I often climbed a hill in the park near my home. This was no ordinary hill; I called it "the Demon" because as you looked down from the top of the hill, there was a near-vertical drop followed by a more gentle slope. In the winter, my friends and I would spend countless hours soaring down the hill at breakneck speed, but most of our time was spent walking up the hill—a trip that was easy at first but became progressively more difficult as we tiptoed up the last few vertical steps. Gains in life expectancy proceed the same way. The reason has to do with a phenomenon in the field of demography known as *entropy in the life table*. Entropy in the life table explains why, despite the news reports, life expectancy for humans will not soon rise to 100 years or higher.

Let's begin with a few facts. Life expectancy in the United States at the turn of the twentieth century was about 45 years—a little higher for women and a little lower for men. Since that time, life expectancy has climbed to about 78 years. This 33-year gain over a 100-year period is the largest and most rapid increase in life expectancy that has ever occurred. Some of the most dramatic and unexpected gains in life expectancy have occurred in the past twenty-five years as death rates from two of the top killers—heart disease and stroke—have declined rapidly. Although death rates have been falling at a progressively faster pace in some countries, the rise in life expectancy has slowed to a crawl. Unraveling this apparent demographic paradox requires coming to grips with entropy in the life table.

At the turn of the twentieth century, 10–15 percent of all babies born in the United States died before reaching their first birthday—principally from infectious diseases. When life expectancy is 45, someone who dies at 50 brings the average up a little bit, and an infant death brings it down. If 10–15 percent of the population is dying as infants, the average comes down a great deal. Improved living conditions and advances in medical technology have led to dramatic declines in mortality among infants, children, and women of child-bearing ages. These improvements not only saved a lot of lives; they also permitted life expectancy to soar during the twentieth century.

Today, far fewer newborns succumb to infectious diseases like influenza, diphtheria, tuberculosis, tetanus, and measles—diseases that preyed on the young early in the twentieth century. In the developed world, less than 1 percent of children now die before their first birthday, and over 93 percent of all babies survive to their 50th birthday or beyond. In fact, death rates at younger ages in most developed countries are so low that life expectancy would increase by only 3.5 years even if every death before age 50 could be magically prevented. This means that the public health initiatives that produced large increases in life expectancy in the twentieth century did so by tapping into a large reservoir of years of life at younger ages. Because death rates at younger ages are so low, this reservoir for life expectancy has been nearly exhausted.

If there is going to be a quantum leap in life expectancy in the future like that observed in the last century, a new reservoir will have to be tapped. The only one that is available in developed nations is that associated with people living to older ages. Exploiting this reservoir will be extremely difficult because survivors to old age have accumulated a lifetime's worth of damage to genes, cells, and tissues—complex damage that poses a huge challenge to biomedical interventions. Furthermore, the risks of death at older ages are so high that the life expectancy reservoir that is available is minuscule by comparison to that available early in the twentieth century. In other words, adding 80 years to the life of an 80-year-old person is far more difficult than adding 80 years to the life of an infant. The implications for life expectancy are obvious. As life expectancy climbs beyond current levels (80 and older), death rates must fall at a progressively faster pace in order to continue to achieve even small increases in life expectancy.

This is the stark reality of entropy in the life table. Increasing life expectancy in a population that is already long-lived is like walking up a hill of increasing slope carrying a stone of increasing mass. Like "the Demon," the higher you climb, the more difficult the remaining climb becomes. Gains in life expectancy are already slowing and entropy in the life table ensures that future progress will be even slower.

In the past, improvements in nutrition, hygiene, sanitation, home environments, and antibiotics proved to be powerful weapons against the infectious and parasitic diseases that killed most people. These external forces of mortality have not been eliminated and it is premature to declare that they have been defeated, but they have been dramatically reduced. We now face what may prove to be an even more difficult force of mortality to control: entropy within ourselves. The age-related causes of death that dominate today, such as heart disease and cancer, arise primarily from biological forces within ourselves—processes of aging that science cannot as yet control.

Adding 33 years to the life expectancy of the U.S. population over the course of the twentieth century by saving the young was an achievement of monumental proportions. This achievement would, however, pale in comparison to the biomedical progress that would be required in order to make the same gain in life expectancy over the next century by extending the lives of older people. It is easy to grab headlines by declaring that life expectancies of 100 years or higher will soon become a reality in developed countries. The inescapable reality of entropy in the life table means that claims of extreme longevity are overly optimistic examples of people who have fallen prey to the Methuselah Trap.

The LAW of MORTALITY

In 1825, a young actuary named Benjamin Gompertz made an important discovery. Actuaries calculate how much money to charge people who buy life insurance. Working at an insurance firm in England, Gompertz was trying to find a simpler and quicker way to calculate the one thing of great interest to a life insurance company: the age at which people are likely to die. Upon gathering mortality information about the people in his part of England, Gompertz detected an interesting pattern. There were a large number of infant deaths, followed by fewer deaths at every age until ages 10–15. After puberty the likelihood of dying increased rapidly, doubling about every ten years until age 80. In other words, he found that a 25-year-old was twice as likely to die in any given year as a 15-year-old; a 35-year-old was twice as likely to die as a 25-year-old; and so on. Gompertz had achieved his goal: he had developed a simple mathematical formula for calculating insurance annuities based on age.

Gompertz then decided to examine the same kind of mortality data for different time periods in England, France, and Sweden. To his amazement, he discovered that the pattern of mortality held. He came to believe that his mathematical formula was far more than just a useful tool for calculating annuities. Gompertz was convinced that he had revealed a fundamental truth about life and death. In fact, he made the bold assertion that he had discovered a law of mortality. He also believed that there was a biological basis for his law of mortality. However, given the limited medical knowledge available at his time, Gompertz could only speculate that the loss of some "vital substance" during the course of life diminished the power of the body to oppose its own destruction—leading to a predictable age pattern of death for individuals in a population. To this day, the mathematical equation that bears his name appears in countless statistical textbooks.

Gompertz had caused nothing less than a profound shift in the way in which age patterns of death are perceived, analyzed, and interpreted. Throughout the remainder of the nineteenth century, scientists vigorously debated whether Gompertz had discovered a "law of mortality" that applied only to mankind, or whether he had revealed a more universal law that applied to all forms of life. If Gompertz's law was both valid and universal, the implications were staggering. A law of mortality implies that there are limits to how long humans and other organisms can live, and biological reasons for why such limits exist.

Raymond Pearl, a biologist from Johns Hopkins University, was one of the first scientists to make a systematic effort to collect data and rigorously test whether Gompertz's law applied to other species. By the early 1920s, Pearl's research had convinced him that a universal law of mortality did exist. However, as he and his colleagues collected mortality data for more species, they were unable to demonstrate that different species have the same age pattern of death—a result that was inconsistent with a universal law of mortality. After nearly sixteen years of continuous research, the highly respected Pearl emphatically declared that a universal law of mortality did not exist.

Pearl recognized that his failure to confirm the existence

of a law of mortality was, in part, due to a problem with his data. In theory, a law of mortality should predict when organisms die from a degradation of the "vital forces" that are intrinsic to the integrity of their own bodies. In the real world, people and other animals die from all sorts of things—accidents, homicide, infectious and parasitic diseases, predators, and suicide—that have little or nothing to do with the intrinsic (aging-related) causes of death that are presumably responsible for a law of mortality.

The problem is that the causes of death that are unrelated or extrinsic to the aging process obscure the age pattern of intrinsic mortality that scientists from Gompertz to Pearl were trying to reveal. In order to solve this problem, researchers would need to identify the cause of death for every animal in their studies—information that Pearl did not collect. Other researchers, in particular those working with small organisms like bacteria or fruit flies or free-living animals like deer or sheep, were faced with the same problem. As a consequence, the search for a law of mortality stopped once Pearl declared failure.

In 1993, the two of us resurrected the search for a law of mortality. Our goal was to find answers to the questions raised by Gompertz and Pearl. Is there a law of mortality that applies to all animals, and why would such a law exist? The reason we thought it was possible to succeed in a task that had eluded so many researchers before us was simple. We had the advantage of hindsight, the benefit of considerable scientific progress made by others, and meticulously collected mortality data for humans and laboratory animals that permitted us to distinguish between intrinsic and extrinsic causes of death.

Modern approaches to the analysis of mortality data permit causes of death to be mathematically eliminated. In our analyses, we eliminated extrinsic (nonaging-related) causes of death: in effect, this was the mathematical equivalent of what improved lifestyles and medical technology have partially accomplished for humans during the twentieth century. Once this was done, the mortality data for the species used in our study exhibited remarkable similarities. The death rates for humans, dogs, and mice dropped to their lowest point around the age of sexual maturity. After puberty, the death rates for the three species increased in accord with the Gompertz equation, and for each species, they were correlated with the length of the reproductive period (the elapsed time between puberty and menopause). In other words, species such as humans and sea turtles that wait for more than a decade before experiencing puberty, and are reproductively active for more than three decades, live longer than mice and dogs, which go through puberty much earlier, and are reproductively active for a shorter duration of time. When the mice, dogs, and humans were compared in our study, the age patterns of mortality for intrinsic causes of death were indistinguishable—just as Pearl had initially predicted. Our study had provided evidence that the law of mortality applies to more than just humans.

When examining the world in which we live, people identify living things by their physical features and colors. A law of mortality implies that living things also possess a characteristic age pattern of death for intrinsic causes. Even more remarkable is the fact that, when compared on a time scale that adjusts for differences in the relative length of their lifespans, different species tend to die at the same ages. For example, people tend to die at between 22,000 and 33,000 days of age (60–90 years), while beagles tend to die at between 3,000 and 5,000 days (8–14 years); in other words, one day of life for a beagle is roughly equal to one week in the life of a person. It is this regularity and predictability of when deaths occur that forms the basis for a law of mortality.

The importance of a law of mortality is that it implies that there are biological reasons for why organisms grow old and why death occurs when it does. A law of mortality also implies that there are limits on how long people can live and, therefore, how high life expectancy can climb without some sort of medical intervention. The intriguing question then becomes whether human ingenuity can find a way to get around the limits on life expectancy that are imposed by the law of mortality.

HAS *the* LAW *of* MORTALITY BEEN BROKEN?

When we compared the mortality experiences of humans, mice, and dogs, the human population had death rates at older ages that were slightly but consistently lower than those observed for either the mouse or dog populations at comparable ages. This was an unexpected result because differences in the age pattern of mortality could not be found when different strains of mice were compared or when mice were compared to dogs. In trying to resolve this discrepancy, we realized that there is an important difference between people and laboratory animals that could explain the lower mortality observed at older ages in humans. Although the laboratory animals were well cared for, they were permitted to die a natural death. In other words, no heroic medical measures were used to extend their lives. Human beings, on the other hand, often seek medical attention when ill, and the sicker we are, the more likely it is that we will seek treatment. The lower death rates for humans suggest that medical technology is extending the lives of people who are living beyond the biological limits implied by the law of mortality—a concept that we refer to as *manufactured time*.

Some prolongevity advocates claim to have already discovered the secret to achieving extreme longevity. The extravagant claims they make are often so far beyond the bounds of scientific credibility that it is easy to expose their false logic and exaggerated promises. But projections of extreme longevity based on hard science can also be mistaken, sometimes with serious consequences. For example, in order to help policymakers prepare for the costs of government entitlement programs like Social Security, researchers have developed mathematical models to make forecasts of life expectancy in the future. Two frequently used methods of forecasting have led some researchers to predict that life expectancy will soon rise to 100 years or higher. Let's take a look at how some well-meaning scientists made these mistaken forecasts.

The RISK FACTOR MODEL

One method that has predicted a life expectancy of 100 years is termed the *risk factor model*. It is based on theoretical estimates of how changes in risk factors (e.g., lifestyles and behaviors such as diet, smoking, exercise, and alcohol consumption) might influence longevity. In 1990, we published an article suggesting that entropy in the life table would make it nearly impossible for life expectancy in developed countries like the United States to rise much beyond 85 years. Soon thereafter, a demographer described our predictions as overly pessimistic and declared that life expectancy would not only reach 85 years, but would soon rise to 100 years. Both research papers received a lot of press coverage. Two teams of scientists looking at the same kind of data came up with dramatically different predictions for the future course of human life expectancy. What is the public supposed to make of this?

Science is and has always been a competition of ideas.

Although researchers may support their ideas with great passion, the scientific process is a dispassionate judge that determines whether new concepts displace old ones or whether confidence in the prevailing concept is bolstered. In either case, scientific progress is made. We examined the assumptions of the risk factor model in order to understand how its proponents could have predicted that life expectancy would rise so much higher than we had predicted.

The health practices of more than two generations of people living in Framingham, Massachusetts, have been followed for decades, providing a wealth of data on cardiovascular diseases and their risk factors. The focus of the investigators who used the risk factor approach was on a subgroup of the Framingham population that they judged to have the healthiest lifestyles of everyone in the study. From these "perfect" people, they further restricted their analyses to the most physically fit thirty-year-olds in the study. In other words, their optimistic predictions about longevity were based upon a highly selected subgroup of people who bear little resemblance to either the vast majority of people in the United States, or even the other people in the Framingham study. The choice of this unusually healthy population was made in order to show the health status and longevity that the investigators believed could be theoretically achieved by everyone.

In addition to data that was strongly biased toward health and longevity, five key assumptions were made that make it easy to see why the risk factor model produced such optimistic predictions. First, it was assumed that everyone in the United States would adopt the idealized lifestyles of the physically fit thirty-year-olds, and that these lifestyles would be maintained by everyone for the duration of their lives. Second, it was assumed that everyone would respond as favorably to these adopted lifestyles as the healthiest thirty-year-old people in the Framingham study. Third, it was assumed that virtually everyone would retain the physiology of a thirty-year-old for their entire lives. Fourth, it was assumed that frailty and disability would be completely and permanently eradicated for all people at every age. Finally, it was assumed that breakthroughs in medical technology would effectively eliminate aging. Although we think these assumptions are biologically impossible to achieve in the real world, they were viewed by the investigators who developed the risk factor model as theoretically achievable by everyone.

With these assumptions, it is a wonder that the risk factor model did not generate predictions of immortality. Studies like this are music to the ears of those who want to believe that extreme longevity can be achieved by adopting "healthy" lifestyles. There is only one problem: utopian worlds without aging and disease where everyone adheres to a perfect lifestyle from birth to death are purely theoretical, akin to the myths and legends of the prolongevity movement, not reality.

Using less biased health and mortality assumptions, scientific studies have demonstrated repeatedly that the death rates for some specific diseases for some people do decline in response to changes in lifestyles. However, these same studies also show that death rates for populations change very little, and the longevity benefits of adopting healthy lifestyles rarely add up to more than a few hundred days on average, and often much less. It is an inescapable reality that every person is genetically unique. This quality makes it impossible for any longevity benefit derived from lifestyle modification to have the same effect on everyone—a biological fact that conflicts with the assumptions of the risk factor model.

Unfortunately, frailty and disability at older ages is the norm, not the exception. As modern medicine continues to manufacture more survival time, there is a growing epidemic of nonfatal chronic conditions among the elderly that is having a profound impact upon the quality of their lives and those who love them—conditions like vision and hearing loss, osteoporosis, osteoarthritis, dementia, and Alzheimer's disease, to name a few. Possibly the most extreme assumption made by some scientists is that medical technology will transform aging into just another disease that can be treated and eliminated—a view that is identical to the discredited philosophies of Bacon, Cornaro, and Brown-Séquard. With this implausible assumption, even some scientists join the ranks of those who advocate extreme prolongevity.

The EXTRAPOLATION MODEL

Another model that predicts life expectancies of 100 years or more is termed the *extrapolation model*. In order to facilitate planning for entitlement programs, the U.S. government requires that forecasts of life expectancy be made for eighty years into the future. The mathematical approaches used for these forecasts can be extremely sophisticated, but one of the most commonly used approaches is quite simple. This approach is predicated on the assumption that a trend in life expectancy observed in the past will continue into the future—an approach referred to as extrapolation. For example, if life expectancy rose by two years from 1980 to 2000, then the simple method would predict a two-year increase in life expectancy between 2000 and 2020. Trends in mortality for all causes of death combined or individual causes of death (such as heart disease and cancer) can be extrapolated in this way. These approaches can be reasonably accurate for short-term forecasts, but they sacrifice realism and accuracy in long-term (eighty-year) forecasts. Scientists using these approaches have predicted that life expectancy will soon rise to 100 years or more—predictions that have not only received public attention but have influenced public policy.

One influential research team has used the extrapolation method to make two predictions that some researchers and government agencies are taking seriously. They predict that death rates will decline by 2 percent at every age, for every cause of death, for every year between now and the year 2080. Their second prediction is that female babies born today in developed countries like the United States and France already have a life expectancy of 100 years.

Why would scientists who know about entropy in the life table generate such overly optimistic estimates for the future course of life expectancy? Death rates declined at younger ages and life expectancy rose throughout the early part of the twentieth century in developed countries like the United States. During the period 1968 to 1992, a new and unexpected mortality trend emerged—death rates from heart disease and stroke declined rapidly for people in middle and older ages. In combination, these mortality trends caused life expectancy to rise more swiftly during the twentieth century than during any other comparable period in history. These unprecedented gains were made during the early stages of a revolution in bio-medical technology. Some scientists have argued, therefore, that death rates will decline even more rapidly in the future because medical advances are likely to occur at an even more accelerated pace. Specifically, these researchers predict that death rates from all causes will decline from the astonishing pace of 1.0–1.5 percent per year observed at older ages from 1968 to 1992, to a 2 percent decline at every age that will continue unabated every year for the next century.

The question is, does the 2 percent assumption, and the life expectancy of 100 derived from it, pass the test of common sense? To begin with, entropy in the life table poses a nearly insurmountable obstacle to the 2 percent assumption. A life expectancy at birth of 100 years requires that almost every cause of death that exists today would have to be reduced dramatically or eliminated altogether. How realistic is that? Even if this assumption were reasonable, it would take a century before life expectancy could climb from 80 to 100 years—a forecast that certainly will not be achieved by children born today. The 2 percent assumption would also require that death rates begin to decline rapidly for diseases that have been rising for most of the past century—primarily cancer.

The extrapolation method does not violate any mathematical principles. It does, however, completely ignore the basic principles of human biology. A simple analogy will make this point clear. Consider how fast men are capable of running one mile. The world record for the one-mile run has been recorded since the middle of the nineteenth century. The record for men has declined steadily from over 5 minutes in 1850 to the current record of 3 minutes, 44 seconds. A simplistic extrapolation model applied to this historical trend would eventually predict that someone will run the mile at speeds that are biomechanically impossible for the human body. The point is that simplistic models can generate predictions that are implausible, particularly when the predictions involve biological phenomena.

Running is an inherently biological phenomenon that is subject to limitations imposed by a genetically determined body design. Similarly, the duration of human life is an inherently biological phenomenon that is subject to limitations imposed by genetic programs for growth, development, and reproduction. If individuals have biological limitations on their duration of life, then in the collective there must also be biological limits on the life expectancy of the populations to which they belong. In order to accurately forecast life expectancy, mathematical models must be based on assumptions which reflect biological constraints that limit the duration of life of every individual, and by extension, the life expectancy of populations. The health and lifestyle choices that we make as individuals must also be guided by an improved understanding and appreciation for the biological forces that influence how, why, and when we age.

The GAMBLER'S ODDS

Several decades ago, the late Nobel laureate Linus Pauling received extensive media coverage when he suggested megadoses of vitamin C as a panacea for a variety of diseases, including genetic ones. Not long ago, headlines around the world broke the news that taking more than 500 milligrams of vitamin C every day may cause genetic damage rather than offer protection from it. The public is bombarded by conflicting information on health like this every day. Part of the problem is that important information is being given to a public that has not been prepared to interpret and understand it. This, in turn, creates an environment ideally suited to exploitation. Prolongevity advocates claim cutting-edge scientific research has proven that vitamins, minerals, hormones, and a host of other substances can combat disease, reverse aging, and delay death. In order to distinguish between fact and fiction at this point, we're going to take a trip to Las Vegas.

On the grand marquee in front of a Las Vegas hotel glittered a sign in bold lettering that read: LOOSEST SLOTS IN TOWN, 98% RETURN ON THE DOLLAR. At the center of the casino was a large bank of slot machines rapidly being fed coins by a huge crowd of people. An outer crowd watched the endless stream of dejected losers walk away with empty coin buckets. Amazing as it may seem, they too were waiting impatiently for their turn to feed the machines. Now and then sirens would shriek, lights would flash, and a lucky winner would walk away with thousands of dollars. Watching this spectacle, it was evident that something was not quite right. With a 98 percent return on investment, why were most of the gamblers walking away dejected at having lost all of their money in a short span of time? With such a large return on the dollar, they should be able to play a long time before running out of money. Perhaps more important, those feeding the machines did not appear to care that something was wrong with the claim made on the marquee in front of the hotel.

Technically, the sign out front was correct. The slot machines are carefully designed to give the casino $2 out of every $100 put into them—the remaining $98 are returned to the gamblers. What the sign failed to advertise is that the money returned would not be evenly distributed. A handful of gamblers win big, while most of the players walk away with thinner wallets and broken dreams.

The mathematics is the same if lottery tickets are substituted for slot machines and state coffers for casino vaults. In the summer of 1998, droves of people flocked to Indiana, Wisconsin, and other states to buy Powerball lottery tickets when the jackpot reached the record level of $292 million. It is, of course, the unequal distribution of reward that drives these games and makes them so appealing. As the jackpot of the slot machine or lottery climbs, more people are willing to invest their hard-earned money even though they know their chances of winning are small. Personal health and longevity are jackpots that are far more precious than money. This explains why so much effort and money has been invested in antiaging substances throughout history, and why people continue to be attracted to the messages pandered by advocates of extreme prolongevity. Unfortunately, the likelihood of striking the longevity jackpot is similarly small.

The unequal distribution of reward and the mathematics behind it is critical to understanding the legitimate benefits that can be reasonably anticipated by modifying lifestyles. To illustrate this point, consider a recent scientific study that involved thousands of women. Half the women in the study were given varying doses of vitamin E and the other half received a harmless placebo. In order to guarantee objectivity, neither the women nor the scientists knew who was given the vitamin and who received the placebo. After completing the study, the scientists reported that the group of women who ingested 250 International Units (IU) of vitamin E every day had 40 percent fewer heart attacks than women taking the placebo. When this finding was reported in the media, people rushed to their local market or health-food store and bought vitamin E.

Although the vitamin E story seems straightforward, it contains logic traps that can cause the unwary to overestimate the health benefits they will receive. For simplicity, imagine that 1,000 women (assumed to be typical of women in the general public) participated in the vitamin E study. From public health statistics, it is known that about 30 percent of all women in the United States will experience the symptoms of heart disease at some point during their lives. For our example, this means that 150 of the 500 women (500 × 0.3 = 150) in each study group (vitamin E and placebo) would be expected to experience heart disease. Rather than a payoff from a slot machine, it is the occurrence of heart disease that is unequally distributed among the women—30 percent are expected to get the disease and 70 percent are not. The 40 percent reduction in heart disease reported by the scientists applies only to the women who would ordinarily experience the disease during their lives.

For our example, this suggests that 60 women (150 × 0.4 = 60) in the vitamin E group avoided heart disease by taking the vitamin. This is excellent news for the 12 percent of women in the vitamin E group (60/500 = 12%) who were fortunate enough to be among those who received the health benefit. However, when applied to women in general, 88 percent of the women who buy vitamin E will not receive the health benefit they thought they were purchasing. Replace vitamin E with other so-called antiaging substances and a similar unequal distribution of benefit will exist. The reality is that substances with reported disease-fighting or antiaging properties will have little or no influence whatsoever on most of the people who take them (exceptions will be discussed later).

The Powerball lottery also illustrates another common logic trap contained within health statistics. In order to win the grand prize, 5 numbers out of an available 48 must be picked successfully, and then 1 number out of 42 determines the winning Powerball number. The odds of winning are published on the back of each ticket—for the grand prize, the odds are 1 in 80.2 million (1/80,200,000 or .000000001). Buying a second ticket doubles the odds of winning the grand prize (2/ 80,200,000 or .000000002). Stated another way, buying two tickets increases your probability of winning the grand prize by 100 percent, investing an additional $100 increases the odds 100-fold (10,000%), and so on. Amazing as it may seem, stories of people investing hundreds and thousands of dollars in these lottery tickets were common during the frenzy to win the $292 million. Clearly, investing in a large number of lottery tickets will dramatically increase your chances of winning. However, the point is that a huge percentage increase in the probability that an extremely rare event will occur is still an extremely small probability.

Medical statistics used to describe the health benefits of vitamins, minerals, and a variety of health practices are often expressed in exactly this way—as a percentage change in the risk or probability of a disease. Classic examples of exaggerated health risks and the misunderstanding of these risks by the public include alarming pronouncements about cancer risks attributed to the use of cell phones and living near power lines. Exaggerated health benefits and the misunderstanding of those benefits include the spectacular reduction in death rates and gains in longevity attributed to taking antioxidants or hormones. Changes in health risks can be declared "significant" by the statistics of science, but the actual magnitude of the cost or benefit is often negligible. Change that is expressed as a percentage can be very deceptive in terms of its actual impact and must be interpreted with great caution. It is always necessary to place benefits and risks into their proper perspective when trying to evaluate the health and longevity consequences of a vitamin, hormone supplement, or behavioral practice.

ADDING HEALTH BENEFITS

Vitamin E has been shown to reduce the risk of heart disease by 40 percent, and selenium reduces the risk of heart disease by 25 percent. Will the risk of heart disease be reduced by 65 percent when these two substances are taken together? Does 500 IU of vitamin E ingested every day provide twice the benefit of 250 IU of vitamin E? The answer to both questions is no; more is not better. Combining substances associated with reductions in the risk of various diseases has an effect which is scarcely greater than that achieved by a single intervention. Changes in mortality risks associated with multiple interventions cannot be added together because most interventions are redundant in their effects on health—once achieved, a health benefit cannot be achieved again. In other words, the whole is less than the sum of the parts. For some people who take a combination of supplements, there may be a reduction in the risk of one or more diseases, but the benefits are always unequally distributed and often far smaller than most expect.

Investing in the personal pursuit of extreme longevity is a gamble worthy of its own grand casino in Las Vegas or Atlantic City. Longevity potions and elixirs are sold today in massive quantities in the United States and elsewhere, earning their proponents billions of dollars along the way. It is unfortunate but inevitably true that the longevity benefits associated with these products are doled out in exactly the same way as slot machines return money to gamblers—unevenly. Perhaps it is no coincidence that the annual meeting of the Society of Anti-Aging Medicine, where many advocates of extreme pro-longevity gather, takes place in the gambling capital of the world—Las Vegas.

MANY PEOPLE CONTINUE TO TAKE vitamin E and other substances with reputed health benefits, and we cannot dispute their reason for doing so. Nobody can know whether they are one of the fortunate few who will receive the health benefit. The potential payoff is so big that even a small chance of winning is better than not playing the game at all. Lots of people are willing to dole out money for that small chance, particularly when the stakes are personal health. The unfortunate reality is that most people who hope to win big in the health and longevity lottery are going to be disappointed.

It was not our goal to depress you by demonstrating that much of what you have been told about aging, disease, and longevity is either false or misleading. To the contrary, the good news is that you can be relieved of the emotional and financial burdens that come from trying to keep up with the latest health fads and their false claims of miraculous health benefits. In later chapters we will demonstrate that there is every reason to believe that current and future progress in science will give rise to technologies and medical interventions that *will* have a profoundly positive impact on your health, quality of life, and perhaps even your longevity.

CHAPTER 4

—⋈—

The Public Health Experiment

*Don't resent growing old. A great many
are denied the privilege.*

UNKNOWN

FROM 1975 TO 1988, a collective groan could be heard
from scientists around the United States as they waited
to hear who among them might receive the annual
Golden Fleece Award. This dubious distinction was often
awarded to scientists performing government-funded
research that Senator William Proxmire from Wisconsin
judged to be a waste of taxpayer money. The award earned
Proxmire national attention and for everyone but the recipient it provided a moment of levity on the evening news.

Several years ago, a young evolutionary biologist by the
name of Michael Rose was conducting research that might
have made him a prime candidate for a Golden Fleece Award.
His research focused on the little flies that buzz around fresh
fruit during the summer, which were collected from a nearby

garbage dump. While most people think of fruit flies as pests and would support research directed toward eliminating them, Rose endeavored to make them live longer through an ingenious experiment. He felt that if the lives of fruit flies could be lengthened experimentally, then the lifespan of humans and other living things could also be increased.

How does one go about changing the basic biology of a species in order to make its members live longer? One way is to mimic the approach that nature has been using for billions of years: evolution by natural selection. Because reproductive success is the primary force that drives evolution, it is possible to alter the biology of a species by controlling which individuals contribute offspring to the next generation, a process known as *selective breeding*. Dog breeders have used this method for countless generations, transforming a single wild dog type into dozens of dog breeds, ranging from the massive Great Dane to the tiny Chihuahua.

Nature has been conducting breeding experiments for billions of years. Every individual of every species has received its genes from an uninterrupted line of ancestors who possessed the genetic characteristics needed to survive and reproduce within the environmental conditions that prevailed during their lifetimes. Human beings are no exception. Individual variation in skin color, eyefold, hair texture, body shape, and even the inheritance of disease—such as Tay-Sachs among Ashkenaze Jews and sickle cell anemia among Africans—are products of the unique environments within which subgroups of the human family arose.

Michael Rose decided to play the role of nature with his fruit flies, hypothesizing that longevity could be extended through selective breeding. This hypothesis was based on theoretical work in the field of evolutionary biology that links reproduction and longevity. To test his hypothesis, Rose permitted only those females that produced eggs later in their lifespan to contribute eggs to the next generation. If repeated over many generations, he predicted that the individuals of each subsequent generation would live longer than the previous one. In human terms, this would be like selecting females aged twenty-five and older to be mothers and then only permitting the daughters who were fertile after age twenty-six to reproduce, and so on for many generations until only women capable of producing babies near the maximum recorded age of menopause would be permitted to be the mothers for the next generation.

The Rose experiment was a success. Each generation of fruit flies lived a little longer than the previous one. As of now, the flies from his ongoing program of selective breeding continue to live progressively longer than their ancestors from the garbage dump. This result is not a fluke; many other scientists have replicated the results of Rose's experiment.

What we learn from this research is that the genetic factors that influence longevity can be modified. Rose believes that if a similar experiment could be performed on humans, a measurable increase in life expectancy would be observed within ten generations (about 250 years). Of course, this prediction will never be confirmed because the selective breeding of humans would be judged unethical and would never be condoned. Rose had to pay a price for his ingenious experiment. The media has nicknamed him, what else, "The Lord of the Flies."

The GREAT LONGEVITY EXPERIMENT on PEOPLE

Countless longevity experiments have been conducted over the centuries, although the enhancement of longevity has rarely been the goal. Mice, dogs, monkeys, birds, flies, worms, and a host of other species have been used in scientific studies for which investigators created carefully controlled living conditions that protected the animals from the hostile forces that exist in a normal environment. Zoo keepers also conduct the equivalent of longevity experiments when they protect their animals from infectious diseases, predators, and environmental extremes. Pets have been the beneficiaries of inadvertent longevity experiments for thousands of years, often receiving better protection from a hostile world than their owners, thereby allowing them to live long enough to experience the effects of aging.

The end result of these longevity experiments is always the same: the animals grow old and die. This may seem like a trivial observation, but in fact it has been an important discovery for researchers interested in aging. Most people do not realize that throughout the history of life on earth, aging has been an exceedingly rare phenomenon. The natural world is, and always has been, a hostile place in which the vast majority of organisms die before they have an opportunity to grow old. The ability to observe the effects of aging on a grand scale is a new phenomenon.

Scientists have learned from these longevity experiments that many forms of life age in a predictable way when protected from the environment. As Gompertz discovered, there is a regularity to the age pattern of death rates within a population—dropping from high levels at birth, to their lowest levels at puberty, after which they pick up speed at an accelerating pace with the passage of time. The changes that most people associate with human aging—bone loss, muscle atrophy, increased risk of cancer and cardiovascular disease, loss of mental function, sensory impairments—occur in other animals as well. These age-related changes may seem as obvious as the inevitability of death, but it is only recently that scientists have begun to recognize the similarity of the aging process across species.

Human beings are in the midst of one of the greatest longevity experiments ever conducted, and nearly everyone is a test subject. Each time an advance has been made in public health and medicine, death rates have declined, and people, on average, live longer. In the past, these changes have included improvements in public sanitation (such as clean water and sewage disposal) that reduced the risk of waterborne and airborne infectious diseases, temperature-controlled indoor living environments that reduced the incidence of respiratory infections, and the development of antibiotics to combat infectious disease—human interventions that we refer to as the First Longevity Revolution.

A larger proportion of the human population is now surviving to older ages than at any other time in the history of our species. For the people who experience a healthful extension of life, this period in the lifespan has the potential to be enormously rewarding. The rewards of a long life go far beyond the individual and extend to family members, friends, co-workers, and a society fortunate enough to benefit from the accumulated wisdom and experience that can accompany a healthy old age. However, for many people, the price paid for living longer has been an extension of old age, an increase in frailty and disability, and a postponement of death that has been emotionally and financially devastating for the dying as well as their surviving relatives. For better and worse, we are now experiencing the consequences of the First Longevity Revolution.

The CHANGING TIDES *of* DEATH

The First Longevity Revolution has brought forth fundamental changes in the way we experience life and death. These changes include, among others, a redistribution of death from the young to the old, a transformation in what causes people to die, and a change in the frequency with which harmful genes exist within the gene pool of our species.

Microorganisms, including those that cause disease, are among the oldest forms of life on earth. These organisms thrived before multicellular life existed; they became an integral part of the cycles of life and death once multicellular life arose; and they will almost certainly continue to thrive when the complex life forms that exist today are long gone. By virtue of short generation times and simple construction, microorganisms reproduce in greater numbers and at a far faster rate than the unfortunate hosts who suffer from the diseases they cause.

Innovations like clean water, sterilized foods, sanitation, waste treatment and disposal, refrigeration, controlled living environments, dental and medical hygiene, improved diets, and antibiotics have given us some control over the infectious and parasitic diseases that have historically taken countless lives, particularly those of children. Children now have a vastly improved chance of surviving to adulthood. Within just a few generations, the highest risk of death has shifted from the young to the old, and infectious and parasitic diseases have been replaced by chronic degenerative diseases as the primary causes of death—diseases that very few people lived long enough to experience in the past.

The First Longevity Revolution has also brought about a less obvious change that involves how common or rare a particular variant of a gene (allele) will be in a population.

Individuals who thrive in a particular environment will naturally produce more offspring than those who do not. As a result, the alleles carried by successful individuals will be more common in the next generation. From generation to generation, there is an ebb and flow in the relative frequency of alleles in a population as luck and the environment influence the rules that govern survival and reproductive success. In its simplest form, evolution can be defined as a change in the frequencies of alleles over time.

In order to have an opportunity to pass on genes to the next generation, an individual must survive to the age of sexual maturity. The infectious and parasitic diseases that have historically taken a heavy death toll among children have therefore had an obvious impact on the genes passed down to the next generation. In addition, there are children born every generation with alleles inherited from one or more parent that compromise their chances of surviving to the age of sexual maturity or their ability to reproduce if they do survive. These harmful genes persist from generation to generation because humans, like many other organisms, have two copies of every gene, and an individual can often get along just fine with only one normal copy of a gene. Also, because a person possesses approximately 100,000 active genes, it is unusual for a single gene to totally determine the survival and reproductive fate of an individual. Natural selection is an imperfect genetic sieve that inevitably allows some harmful genes to leak through to the next generation by hitching a ride with the many genes that have either a neutral or beneficial effect on the survival and reproductive success of the individual.

There are exceptions to the rule that a single gene does not determine the fate of an individual. Some deadly disorders, such as phenylketonuria (PKU—the inability to produce a single protein required to sustain life), early onset diabetes, hemophilia, and childhood leukemia, are caused by a single lethal gene. Today, children born with these disease-causing genes are often able to live long and healthy lives—a great triumph of modern medicine and a source of joy for the parents of these children.

Improved public health measures and medical technology have thus circumvented the forces of natural selection that have shaped the human gene pool by permitting individuals to survive and reproduce who would have been denied this opportunity in the past. This is not a value judgment. It is a simple statement of fact. Some may argue that the gene pool is being harmed while others may revel in the powers of modern science. History has already proven that making value judgments about the relative "worthiness" of individuals based on some arbitrarily chosen physical, mental, or genetic characteristic is a destructive practice with no redeeming value. The ethics of technology can be debated, but its progress cannot be abated. It is only through education and broad understanding that safeguards can be established to make sure that scientific progress is indeed progress. The challenge for the twenty-first century will be to learn how to deal with the consequences of wielding the power to modify the human gene pool.

POPULATION AGING

The First Longevity Revolution has had a profound impact on the shape of human society. The age composition of a population (referred to as "age structure" by demographers) reflects historical trends in birth rates and death rates. For most of the 130,000-year history of anatomically modern people, the age structure of our species has been shaped like a pyramid, with a broad base that represents a large number of children. If this age pyramid could be extended to include the nine-month period between conception and birth, the base would be even broader, because it would include all the embryos and fetuses that are lost prior to birth. The pyramid tapers toward a small apex that represents the few individuals who survive to extreme old age. Until recently, the pyramidal age structure of humans was just like that of other animals—a large number of young diminishing to a handful of individuals from earlier generations who managed to survive to older ages.

After World War II, the shape of this age structure began to change. At first, a large bulge appeared at the base of the pyramid because of unusually high birth rates from 1946 through 1964. This bulge, or baby boom, contributed to the rapid population growth that was observed around the world in the twentieth century. When the baby boom slowed down after 1964, an indentation that lasted for about a decade appeared at the base of the pyramid. As death rates at middle and older ages declined rapidly during the twentieth century, more people managed to survive to older ages—people who in earlier times would have died. The result of these demographic changes is an age structure that has probably never been seen before for any species—an age structure that has shifted from a pyramid to a rectangle. Although this demographic transformation—known as population aging—may seem like something of interest only to population scientists, it is a harbinger of unprecedented social and biomedical changes that will have a profound impact on the lives of virtually everyone.

When the Social Security program in the United States was created in 1935, its designers estimated that the number of people who would be able to draw benefits would never exceed 25 million. Their estimate was based on the premise that the age structure of the U.S. population would retain the usual pyramidal shape. The planners of Social Security could not foresee the unprecedented demographic transformation that was coming. By the year 2000, just sixty-five years after these initial projections, Social Security was responsible for over 38 million beneficiaries. The baby boomers born after World War II will begin to reach retirement ages in the year 2011, and when they do, the number of beneficiaries will quickly soar to 70 million or more. Unanticipated population aging will also have a profound impact on the financial integrity of age-based entitlement programs like Medicare. In addition to the financial challenges, population aging will force changes within many of our social institutions, such as work, retirement, and perhaps even marriage.

Population aging at the global level caught almost everyone off guard. All eyes were focused on population growth and the dire consequences of a human population that appeared to be expanding out of control. In an article entitled "The Aging of the Human Species," published in *Scientific American* in 1993, we predicted that population aging would soon replace population growth as one of the most important concerns of pubic policy—a prediction that has become a reality. Because the biomedical war against disease and death is accelerating at an ever faster pace, the consequences and importance of population aging will also accelerate over the coming decades. In reality, we are only beginning to comprehend what science has set in motion by modifying the forces of natural selection that have shaped the evolution of human aging.

The CURRENT STATUS *of the* FIRST LONGEVITY REVOLUTION

For most of the past 130,000 years, life expectancy for human beings was probably no more than an average of about 20 years. A hostile world molded the lives of our ancestors and forged the genetic structure that defines our species. The slow rise in life expectancy during most of the last two thousand years and the rapid rise throughout the twentieth century occurred because we learned how to partially subdue our environment. The First Longevity Revolution has been a protracted war against the many external forces of nature that have historically challenged human survival.

In developing nations, the First Longevity Revolution is still in full swing. In India, parts of China and Southeast Asia, and in most parts of Africa, large and rapid increases in life expectancy could be achieved today through concerted efforts to control the forces that cause high infant and child mortality. Programs designed to reduce malnutrition or to immunize children against infectious diseases could save millions of children who now die from diseases that are entirely preventible. Although developing nations would benefit from programs designed to reduce death rates at any age, no approach will lead to more rapid gains in life expectancy than programs targeted toward saving the lives of children.

Even in developed regions of the world like North America, Western Europe, Australia, and Japan, the benefits of the First Longevity Revolution have yet to be fully realized. Realistic reductions in mortality will continue to add years to the life expectancy of these nations. However, a redistribution of death from the young to the old can happen only once for a population. Large reductions in death rates may still be achievable in developed countries, but entropy in the life table means that large and rapid increases in life expectancy are highly unlikely. We have stopped dying young, but we still die when we become old. Changes in life expectancy simply reflect the fact that many more of us now have the opportunity to grow old. Subsequent gains in life expectancy will be harder to come by, and will require profound and difficult changes in human behavior, society, and biomedical technology.

UNDER THE MOST OPTIMISTIC SCENARIOS, the First Longevity Revolution may be able to achieve mortality reductions that result in a life expectancy of about 88 years for women and 82 years for men, with an average for both sexes combined of 85 years. The sorts of changes that would be required to achieve these goals would include: (1) significant reductions in the embarrassingly high levels of infant mortality that still exist among some subgroups of the population; (2) immunizations for all children; (3) elimination of smoking and the excessive consumption of alcohol; (4) elimination of obesity through changes in dietary habits; (5) regular exercise; (6) ubiquitous use of seat belts in automobiles; (7) safe sexual practices by all; and (8) access to medical insurance and basic medical care for every member of the population. This utopian world is not realistically achievable; but further movement in this direction could yield mortality reductions that are large enough to raise life expectancy at birth an extraordinary eight to ten years beyond that observed today. Such a gain in life expectancy would not equal those experienced earlier in the twentieth century, but it would still be a remarkable achievement.

By extending survival into old age for unprecedented numbers of people, the First Longevity Revolution has been a success of monumental proportions. However, as much as we might wish it were not so, the biological structure and organization of living things imposes limits on their physical and physiological capabilities. Since longevity is an attribute of living things, it must have limits as well. The wars waged against external forces of mortality in the First Longevity Revolution will never truly end. But human knowledge has already progressed to the point where it has begun another longevity revolution. In this second revolution, science will be used to manipulate the genetic instructions that determine the structure and function of the internal processes which give biological meaning to life and place limits on its duration.

Chapter 5

—⋈—

Manufactured Time

Live each day as if it were your last—
someday you'll be right.

KATHLEEN NORRIS

IN 1970, A FIFTY-YEAR-OLD MAN by the name of Mark from Duncan, Oklahoma, was taking his usual morning stroll to work. Every day after arriving at his office building, Mark would run up the three flights of stairs to his office rather than use the elevator. Although many years had passed, Mark still thought of himself as an air force sergeant in the South Pacific during World War II. His exercise ritual was a daily affirmation that he had not lost the vigor of his youth. In his five decades of life, Mark had never visited a doctor or hospital for any significant health problem. His routine on this early December morning was no different than it had always been. However, on this day Mark noticed a pain in his chest after his run up the stairs—a pain that he did not recognize. Not knowing what to make of it, he walked back down the stairs and ran up the three flights again. This time, the pain took his breath away.

Duncan is a small town, and Mark's family physician, a personal friend, diagnosed the problem easily and quickly. Mark was told that he needed to see a heart specialist, and if he did not go immediately, he probably would not live to see the new year. Mark was a stubborn man, a Christian man, and a man intensely devoted to his family. His children were coming home from college for what Mark considered the most important day of the year: Christmas. Nothing was going to ruin the family celebration—not his health, not his doctor, and especially not a hospital. The heart specialists were just going to have to wait.

The day after Christmas, Mark was placed in a wheelchair and put on an airplane bound for Houston, Texas, where he met with the famous surgical team of Debakey and Peterson— the pioneers of coronary bypass surgery. A few days later, Dr. Peterson opened up Mark's chest and grafted one new artery onto his heart, making him the one hundred eighty-first coronary bypass patient in the world. Within a few days, Mark was forced to get up and walk around and then return to his room, where he chatted with his roommate who was recovering from the same surgery—Andy from the famous radio show *Amos 'n' Andy*. In a remarkably short time, Mark was returned to his family in excellent health.

Mark had always been a physical man and disciplined almost to a fault. He quickly dropped his weight from 230 to 175 pounds, watched his diet and alcohol consumption, swam regularly, and jogged along the cattle pastures near his home almost every day. Mark lived another twenty-eight years, sharing many a joyful Christmas with his family. Even near the end of his life while fighting a losing battle with Alzheimer's disease, Mark was able to care for his beloved wife, who was losing her own personal battle with emphysema. My dad, Mark Carnes, was a remarkable man, and his friends and family are thankful for the miracle of medical technology that was able to *manufacture* a healthy quarter century of life that we were privileged to share with him.

In the summer of 1974, a young man twenty years of age was nearing the end of a two-week road trip through the western United States and Canada. On the day before arriving home, he felt something funny on the car seat and reached down to see if he was sitting on a marble or some change that might have fallen out of his pocket. Nothing was there. That evening, he discovered a large, tender lump at the base of his spine that was growing more painful by the minute. He went to the doctor the moment he arrived home.

His doctor told him that he was born with a congenital malformation that resulted in a hole at the base of his spine, a condition referred to as a pilonidal cyst. The hole was so small that his mother had never detected it during infancy. Sometime after puberty, hair had grown into the hole, and eventually an infection developed. Left untreated, the infection could have worked its way into the intestines or abdominal cavity and killed the young man.

A few days later, a surgeon removed the cyst. Although a routine procedure today, this kind of surgery was not available in the past, often making congenital problems like this a lethal condition. Recovery from this surgery was difficult, but by six months the young Jay Olshansky had recuperated well enough to bound up a flight of stairs with ease. Today, I still run upstairs and appreciate now more than ever how medical technology has *manufactured* survival time for millions of people like me who would have died at a much younger age in the absence of medical intervention.

At age thirty-three, Neil, a professor at Yale University, was enjoying the prime of his personal and professional life. One night he was dancing with his wife at a wedding when he felt a pain in his chest. The pain subsided when he sat down but returned a few days later during a game of squash. After an EKG, Neil was given a clean bill of health. "It's psychological," he was told. A stress test was scheduled in two weeks. A week later, while driving home with his wife, Grace, Neil experienced chest pains that were even more severe than before, accompanied by a cold sweat and shortness of breath. This time he knew what was happening: he was having a heart attack and he needed to get to a hospital fast. Grace could not drive a stick shift well, so Neil drove himself to a hospital located twenty minutes away—running red lights and driving on sidewalks and over curbs to pass cars along the way. Upon arriving at the emergency room, Neil said in his usual polite tone, "Excuse me, I'm having a heart attack." Ninety seconds later, his heart stopped. It took another minute before a defibrillator could successfully shock the heart back to life.

Seven days later, a tiny balloon was placed into one of Neil's clogged coronary arteries and inflated. This procedure, known as balloon angioplasty, compresses the plaque against the wall of the artery and improves blood flow. In the years that followed, Neil changed his diet, began to exercise, and started a family. At the age of forty, he required triple bypass surgery. Once again, medical technology saved his life. Neil is now forty-six years old. His career is flourishing and he is enjoying a healthy life, along with Grace and their two children.

The physicians and other health-care workers who provide medical care for people with disease, and the biomedical researchers who study diseases in order to develop more effective identification, treatment, and prevention, are becoming medical miraclemakers. As these three examples illustrate, dramatic progress has been made toward understanding and treating diseases and disorders that have historically been life-threatening. The importance and magnitude of this progress and its impact on life expectancy should not be underestimated. In developed countries where most people die beyond the age of sixty-five, medical miraclemakers are the primary reason why death rates are declining, life expectancy is rising, and survival rates for most lethal diseases are improving so dramatically. It is likely that most of you have had, or will have, your life saved at least once by the new technologies of medical science.

Manufacturing Survival Time

The prolongation of life has always been a goal of biomedical science. The First Longevity Revolution gave more people a chance to achieve their longevity potential by reducing threats to health and longevity that originate from outside the body. The medical miraclemakers of today have begun treating the symptoms of age-related diseases that originate within the body, and they have begun the challenging task of seeking cures for the diseases that are a genetic legacy from our ancestors. The technologies of modern medicine (surgery, dialysis machines, respiratory aids, to name a few) already permit many people to survive beyond their inherent longevity potential.

The three men in the stories that opened this chapter lived longer because medical technology was able to *manufacture* survival time for them. We created the term *manufactured time* to describe the additional days lived as the result of a successful medical intervention for a life-threatening condition. The term implies that death would have occurred without the intervention. Of course, the actual amount of survival time that is manufactured cannot be measured with precision because it is not possible to know exactly how long an individual might have lived in the absence of medical intervention. Despite this problem, manufactured time is a useful concept because it emphasizes the point that many people are already living beyond the longevity potential with which they were born.

Recall that Benjamin Gompertz, the British actuary, detected a common age pattern of mortality for humans, and published a mathematical description for this age pattern that he called a law of mortality. Our research revealed a remarkably similar age pattern of death for species as different as the mouse, dog, and human. Upon closer inspection, we discovered that people had lower death rates at older ages than the laboratory animals. This observation surprised us because we had expected no mortality differences between the species. We theorized that the lower death rates for humans were the result of the cumulative effect of *manufactured* survival time— medical interventions that were not made available to the laboratory animals. This led us to the startling conjecture that medical science may have already pushed human survival beyond the limits implied by a law of mortality.

The DAWN *of* EVOLUTIONARY MEDICINE

Why, in a body of such exquisite design, are there a
thousand flaws and frailties that make us vulnerable to
disease? If evolution by natural selection can shape
sophisticated mechanisms such as the eye, heart, and
brain, why hasn't it shaped ways to prevent nearsight-
edness, heart attacks, and Alzheimer's disease? If we
can live a hundred years, why not two hundred?

RANDY NESSE AND GEORGE WILLIAMS,
WHY WE GET SICK

The human body is a miraculous living machine, with inter-
acting components operating at a level of complexity far
greater than any mechanical device ever created—from the
DNA that orchestrates the symphony of life to a brain capable
of storing, sorting, and interpreting massive amounts of complex information. However, the same living machine that
evokes wonder and amazement also gives rise to the flaws and
frailties noted in the quote above by Nesse and Williams. In
Why We Get Sick, the authors suggest that an evolutionary
understanding of disease can give rise to new approaches to
its management and treatment. This evolutionary approach to
medicine emphasizes the need to develop an understanding
and appreciation of the natural healing mechanisms of the
body, knowledge that is particularly relevant to the treatment
of the diseases of aging.

Taking an evolutionary approach to medicine raises many
questions about our traditional responses to the symptoms of
disease. For example, when a microorganism invades the
body, the body responds by automatically raising body temperature. Most people immediately reach for aspirin or some

other medication that will reduce the fever as quickly as possible. A fever may be an unpleasant experience for people, but it is lethal for the invading organism. While fever-reducing medications make us feel better, they negate one of the defensive responses to invasion made by the body, possibly prolonging the disease or even worsening the illness.

Similarly, morning sickness can be a stressful ordeal for many women during the first three months of pregnancy. Fearing an abnormal pregnancy, these women often seek medical advice, and when the symptoms are severe, they may take medication to control the nausea. According to Nesse and Williams, morning sickness is an evolutionary adaptation. The first trimester of pregnancy is a vulnerable stage in the development of the fetus because of the incredible amount of biological activity that is taking place. Nausea, vomiting, and aversions to food may be adaptive responses designed to minimize exposure of the fetus to potentially damaging toxins during this critical phase of development. It may not be wise to interfere with a normal response designed to protect a developing fetus. These lessons from evolutionary medicine emphasize the importance of respecting, protecting, and enhancing the natural healing and protective powers of the human body.

In recent years, antibiotic-resistant strains of tuberculosis and meningitis have appeared in the United States and elsewhere—diseases that were previously thought to have been conquered. The emergence of new diseases and the reemergence of more virulent strains of known diseases is a serious health threat that has already received public attention through popular books and movies (such as *The Hot Zone*, *The Coming Plague*, and the 1995 movie *Outbreak*). The traditional response to these threats has been the development of new and even more powerful antibiotics. In military terms, this strategy can be likened to an escalating arms race. When a targeted microbe is not totally eliminated, progressively more potent weapons are required to combat the resistant survivors of the previous battle. Unfortunately, this battle has already given rise to microbes that are resistant to all known antibiotics, a problem that has the medical community seriously alarmed. Lessons learned from evolutionary biology suggest an alternative approach to the health risks posed by microbial adaptation. First, doctors must stop the indiscriminate use of antibiotics. As with alternating pesticides in agriculture, we can make it difficult for microbes to adapt by prescribing different kinds of antibiotics on a rotating schedule. Second, we can take advantage of the rapid adaptation of microbes by using antibiotics designed to lure microbes toward adaptations that make them vulnerable to other antibiotics.

Evolutionary medicine is based on a fundamental respect for the ability of the body to protect and heal itself. This concept is gaining support among the practitioners of Western medicine and their patients. Chiropractors used to be thought of as fringe elements in the world of medicine. Today, their documented contributions to public health have earned them both acceptance and respectability. Doctors and scientists are now investigating cancer treatments that encourage the immune system to attack cancer cells as if they were infectious microbes, an approach consistent with the principles of evolutionary medicine. As with anything new, it will take time before these and other such treatments fully make their way into mainstream medicine. Efforts are currently underway in the United States and elsewhere to subject the methods of evolutionary medicine to the time-honored methods of scientific testing. Those methods that prove to be effective in the treatment of disease will then become acceptable tools for maintaining health and extending longevity.

DESIGN FLAWS *and* ODDITIES

The human body is a miraculous biological machine, with a remarkable ability to protect and repair itself. Although evolution is not directed toward perfection, the countless wonders of the living world invariably bring images of perfection to mind. This notion has led some to portray health and extreme longevity as normal, and sickness, aging, and even death as abnormal departures from a normal healthy state. It would follow from this logic that sickness and aging can be reversed because they are caused by either external events or individual lifestyle choices.

A closer examination of the living world reveals that this is not the case, as a few examples relevant to human aging will illustrate. The ability to walk upright may have been a key evolutionary adaptation for early mankind, but it also causes a variety of health problems, particularly among older people. Over time, vertebrae that are aligned in the same direction as the force of gravity can become compressed, leading to lower back pain, pinched nerves, and ruptured disks. The human appendix originally housed bacteria used to digest plant matter. Today, the appendix is primarily a source of infection that we know as appendicitis. Other design flaws include a useless vestigial tail (the coccyx), a digestive and respiratory system that cross in the back of the throat, often leading to choking, and an inverted attachment of the optic nerve in the eye that creates a blind spot and increases the risk of a detached retina. The response of the body to puncture wounds is to conserve fluids by constricting the cardiovascular system, a life-saving adaptation in the dangerous world of our ancient ancestors. Today, this adaptation leads to an elevated risk of hypertension and stroke among people who survive to older ages.

These and other human design flaws are described in detail in a book appropriately entitled *The Scars of Evolution* (1990), by Elaine Morgan. Similar imperfections, contraptions, and evolutionary relics in plants and animals are discussed in *The Panda's Thumb* (1980), by Stephen Jay Gould. Gould makes the interesting point that these "oddities of nature" provide some of the strongest evidence for evolution because an "omnipotent creator" could have achieved perfection—an argument also recognized by Charles Darwin.

Oddities and poorly designed anatomical structures exist because evolution does not strive for perfection; in fact, it does not strive for anything. Natural selection simply works with whatever biological material is available at the moment. Variations in biological design that confer a reproductive advantage within the prevailing environment will increase at the expense of those that do not. Recognizing that people are a hodgepodge of biological design features that would make Rube Goldberg proud leads to a much different perspective on aging, health, and longevity than one based on departures from expected perfection. Most of the design flaws in the human body would not have been a serious problem for our early ancestors because of their short life expectancy, but they can and do lead to significant health problems for people living well into old age today. In truth, sickness, aging, and death make perfect sense for living machines with body designs that were never intended to be tested in the laboratory of extended life. Accepting this reality opens new approaches to the treatment of disease.

BLOOD BANK *of the* FUTURE

One exciting prospect for using the body's own systems for healing comes from the umbilical cord, which is usually discarded after birth. Scientists have recently discovered that the blood in the umbilical cord is a rich source of special cells known as stem cells. Stem cells are special because they have the ability not only to replace themselves, but also to grow into different kinds of specialized cells, such as red blood cells, white blood cells, and platelets. Stem cells can transform themselves because they have not undergone the genetic changes (turning genes either on or off) that occur when a cell becomes specialized through a process known as differentiation. Once a cell becomes differentiated (for example, a nerve cell), it is less able to repair damage that accumulates in its DNA, the sort of damage often associated with the aging process.

Although the potential therapeutic value of stem cells has been appreciated for some time, physicians did not start harvesting and using cord blood to treat human diseases until 1988. Cord blood has already been used successfully in the treatment of anemia, immune disorders, breast cancer, lymphoma, neuroblastoma, leukemia, multiple sclerosis, and numerous genetic diseases. For example, patients who undergo aggressive radiation therapies for cancer often suffer a significant loss of the bone marrow that produces blood cells. The traditional response—a bone marrow transplant—is both difficult and dangerous. An injection of cord blood is far less invasive than a marrow transplant, and recent studies have demonstrated that the injections are just as effective. Once injected, the stem cells migrate to the bone marrow, where they begin the task of producing healthy blood cells. Researchers who study cord blood believe that it could also play an important role in future treatments for osteoporosis, AIDS, Hodgkin's disease, cystic fibrosis, and a variety of cancers.

Currently, supplies of cord blood depend on donors who allow the cord blood of their child to be made available to others. As the demands and applications for cord blood grow, the harvesting, freezing, and storage of cord blood in blood banks may become routine. The huge cost savings associated with using cord blood injections as an alternative to bone marrow transplants will no doubt encourage their use and they may even be required by insurance companies in the future. A better understanding of how the stem cells of cord blood resist the genetic consequences of aging may eventually contribute to the development of new and improved treatments for many diseases and disorders associated with aging.

Cultivating *the* Primordial Cells *of* Life

A related technology that will soon be used to treat aging-related diseases involves specialized cells that exist for only a brief moment in the earliest stages of life. When a fertilized egg begins to divide, a long developmental process begins that eventually culminates in the construction of a human being. During one of the earliest stages of this process, a hollow sphere of cells known as the blastula begins to form. Within the blastula is a small cluster of cells known as embryonic stem cells. These embryonic stem cells will eventually differentiate into the specialized cells that form the various tissues and organs of the developing embryo, including the embryonic germ cells that will give rise to eggs and sperm. Normally, once an embryonic stem cell differentiates, it is permanently committed to the developmental pathway that forms a specific type of tissue.

In the fall of 1998, an astonishing scientific discovery was announced that went largely unnoticed by the general public. Incredibly, Dr. James A. Thomson of the University of Wisconsin and Dr. John Gearhart from Johns Hopkins University had figured out how to isolate embryonic germ cells and cause them to revert back to embryonic stem cells. Stem cells can be grown and maintained in the laboratory, and because they are precursors to other kinds of cells, they have the potential to become virtually any kind of cell that exists in the body. This discovery raises the amazing prospect that in the future, critically ill patients may not need to wait for a donor in order to receive an organ transplant—the needed organ or tissue will be grown in the laboratory. More immediately, stem cells injected into a diseased organ such as the heart, liver, or brain could be induced to replace damaged or compromised tissue within these organs. Although the possibilities are endless, diseases like insulin-dependent diabetes, Alzheimer's, Parkinson's, and atherosclerosis are considered prime candidates for this future technology. This kind of advance has the potential to manufacture survival time by curing the diseases of aging rather than simply treating their symptoms—a biomedical breakthrough that could have a profound impact on the way medicine is practiced in the twenty-first century.

Unfortunately, the source of embryonic germ cells is somewhat problematic. At present, the blastocysts that harbor these cells come from embryos that are either aborted (which are difficult to capture at the proper developmental stage), or donated after being created by in vitro fertilization. Both sources give rise to an ethical dilemma that remains unresolved: what to do with a technology that must terminate a potential life in order to save or extend an existing life.

This technology also poses technical problems that have not yet been solved. As with organ transplants, tissue rejection forms a serious obstacle to the transfer of embryonic stem cells. Until this problem is solved, the best source of embryonic stem cells would be the patient; but these cells are not available to anyone alive today because they only exist for a brief period of time during the very early stages of a developing embryo. In theory, these cells could be created by cloning an embryo from the patient's own cells. However, it is an understatement to say that there will be ethical dilemmas associated with cloning an embryo solely for the "harvest" of its stem cells. Scientists also face a more fundamental obstacle to the use of this technology; it is not yet known how to direct embryonic stem cells to make a specific kind of cell, tissue, or organ. Nobody knows how long it will take to resolve these ethical and technical problems. One thing is for sure: when they are resolved, the future of medicine will be changed forever.

LIVING BEYOND OUR BIOLOGICAL LIMITS

Probably everybody reading this book knows someone who would be dead were it not for a life-saving medical intervention. In the developed countries of the world, there are enough people living on manufactured time to have a measurable impact on the life expectancy of the country in which they live.

In North America, Western Europe, Japan, and Australia, life expectancy increased dramatically during the twentieth century. Population scientists attribute these gains to improvements in sanitation and living conditions rather than to advances in medical technology because the bulk of the increase occurred well before the introduction of antibiotics. Over the last twenty-five years, death rates have continued to decline rapidly throughout the developed nations of the world, particularly among people at middle and older ages.

Some scientists credit the technologies of biomedical science for these declines, while others argue that the adoption of healthier lifestyles is responsible. Whatever the cause, the declines observed in death rates in recent years have inspired some advocates of prolongevity to make the bold prediction that most fatal diseases will eventually be eliminated and that life expectancy will continue its dramatic rise in the future. The most zealous advocates of prolongevity totally reject the existence of biologically imposed upper limits on lifespan (individual longevity potential), declaring that the life expectancies of human populations will climb indefinitely. For some, the recent declines in death rates at older ages prove that even if there is a biological limit to life expectancy, it must exist somewhere beyond the visible longevity horizon.

We suggest that the reason prolongevists cannot detect a biological limit to life expectancy is because that limit has already effectively been surpassed in some populations. Debates over limits to life expectancy may sound like an intellectual game, but this issue has profound implications for science and public policy. Assume for the moment that all of the major fatal diseases are a product of hazardous environments and poor lifestyles and may therefore ultimately be eliminated. If so, then biomedical research should focus on identifying those environments and lifestyles that cause disease, and then find ways to protect people from harmful environments and encourage them to adopt healthier lifestyles. If this view determined public health policy, then the biomedical community would try to treat the symptoms of a disease only until the environmental and behavioral causes for the disease could be identified and controlled. In fact, this view has motivated much scientific research.

Now assume that for human populations buffered from the hazards of nature, many people are already living beyond their longevity potential because of medical intervention. This means that the observed life expectancy of the population to which they belong has been increased artificially by the addition of *manufactured* survival time. When we acknowledge the internal, biological sources of disease and aging, it becomes clear that future improvements in lifestyles and the environment will only produce modest increases in life expectancy in countries where people already live to old age. Additional significant increases in life expectancy can only come from advances in biomedical technology that alter the course of aging itself. The medical miraclemakers will have to manufacture progressively more survival time for progressively more people surviving to progressively older ages in order to continue the decline in death rates. Entropy in the life table and a genetic legacy of design flaws and intrinsic disease processes received from our ancestors will make this a daunting task. In other words, doctors and scientists will not only have to deal with the symptoms of disease, they will have to attack the underlying biological processes that cause aging and disease. If successful, these biomedical breakthroughs will lower or remove biological barriers that currently deny access to extreme longevity for most people.

OVER THE COURSE of the twentieth century, our species has experienced a dramatic shift in the primary source of threats to life—from external environmental threats to the life-threatening internal failures of a human body that was not designed for extended survival. This unprecedented mortality transition has been the direct product of an avid pursuit of knowledge about basic biology and the environment that gives it meaning. The biomedical response to health crises has now begun to shift from medical care intended to treat the symptoms of illness to interventions that are designed to cure the underlying causes of disease. Although great strides have been made in understanding the remarkable capacity of the human body to heal itself, almost everyone eventually confronts a health crisis that cannot be overcome without help. Every threat to life that is averted by a medical intervention artificially extends the lifespan of an individual and generates manufactured survival time that incrementally raises the life expectancy of the population to which that individual belongs.

The medical miraclemakers of the twentieth century have done an extraordinary job of saving the lives of people at every age, often manufacturing survival time more than once for the same person. It is easy to appreciate the big miracles, but we also benefit from countless small miracles performed by those in health-related occupations every day. From obscure scientists making important biomedical discoveries, to family physicians detecting potential health crises before they become a problem, to dentists and dental hygienists preserving our teeth, to pharmacists dispensing miracle drugs, our lives are qualitatively better and considerably longer on average than was the case just a few generations ago. These facts should be a reminder of how good life is today.

The miraclemakers of the future will undoubtedly continue to discover new ways to prolong life. However, as the population continues to age in the twenty-first century, we must balance the survival time that is manufactured with an equal emphasis on the quality of that survival time. This will require an entirely new approach to biomedical technology— an approach that moves beyond the treatment of diseases and their symptoms to a determined effort to identify and modify the mechanisms responsible for aging and disease.

Chapter 6

—⋈—

Antioxidants

*Of all the wild chimeras which have in all ages
haunted the minds of fantastic dreamers, none has
taken so great a hold on its votaries as the search for the
Elixir Vitae or means of prolonging human life beyond
its allotted span.*

Edmund Goldsmid

WHEN THE UNITED STATES detonated atomic bombs above the cities of Hiroshima and Nagasaki to end the war with Japan during World War II, everything and everyone in the immediate vicinity of the detonation was incinerated. Those who were farther away from the epicenter of the blast were exposed to a short burst of radiation that varied in intensity according to their distance from the blast and whether they were indoors or outdoors.

At the time of the bombings, the long-term health consequences of exposure to nonlethal doses of radiation were poorly understood. Through a joint treaty between the United

States and Japan, the atomic bomb survivors became the subjects of an intense biomedical study that continues to this day. The subsequent development, construction, and testing of nuclear weapons during the Cold War exposed even more people to varying doses and intensities of radiation. Although peacetime uses for radiation existed long before the bomb, the proliferation of nuclear technologies after the bomb—including such areas as treatments for cancer and the generation of electricity with nuclear power—has further increased the need to understand the health consequences of exposure to radiation.

RADIATION *and* AGING

Radiation generated by the sun and radioactive rocks in the environment has been an important health hazard throughout the history of life on earth. As a result, every species from simple organisms like bacteria to complex animals like humans has developed a remarkable ability to protect against and repair the kind of biological damage caused by radiation. The reason for discussing radiation in a book about aging is that some of the genetic damage caused by radiation is identical to that postulated in the free radical hypothesis of aging. The similarities to aging at the cellular level led scientists of the past to describe the biological consequences of exposure to radiation as *accelerated aging*—a term that radiation biologists no longer view as accurate. Nonetheless, the similarities between damage caused by radiation and free radicals and the substances used to protect against this damage (antioxidants and radio-protectors) have interesting consequences for the battle against aging.

For years, prolongevity advocates have championed the

antioxidant properties of a variety of vitamins, minerals, chemicals, and foods as a way to delay or reverse the aging process. Interestingly, decades before antioxidants became a health craze, researchers working for the military were striving to create man-made compounds that would be more effective antioxidants than those produced within the body or contained within the foods we eat. After World War II, military strategists could not ignore the possibility that troops would be fighting a nuclear war. Along with other protective measures, compounds with enhanced antioxidant properties were sought to lengthen the time that soldiers could perform combat duties in a battlefield contaminated by radiation. The antioxidant compounds that the military scientists created to protect soldiers from the effects of exposure to radiation are termed *radio-protectors*.

As the threat of nuclear war has waned, radio-protector compounds have been released to scientists who are not affiliated with the military in order to determine whether they have biomedical applications. The knowledge that has been gained about how these compounds reduce the genetic damage induced by radiation has considerable relevance to the potential antiaging benefits of vitamin supplements like C and E.

One of the more effective radio-protectors developed by the military was designated WR-2721, the WR indicating that it was developed at Walter Reed Hospital in Bethesda, Maryland. David Grdina, a professor of radiation biology at the University of Chicago and one of the world's leading authorities on radio-protectors, has launched studies on the use of this compound to protect cancer patients from the damaging effects of radiation therapy and to reduce the risk of DNA mutations caused by drugs like AZT that are administered to AIDS patients. His research has already confirmed the protective effects of WR-2721 in laboratory animals exposed to radiation, and it is now being tested in clinical trials to verify its effectiveness and safety for use by people.

Just as we have predictable times of the day when we eat, sleep, work, and perform other activities, the cells of the body have an activity cycle divided into time periods when specific functions are performed. This organization of cellular activity is known as the cell cycle. A critical activity that takes place during one period of the cell cycle is the identification and repair of damage to DNA. Grdina believes that WR-2721 lengthens the cell cycle, which is analogous to making a day longer than twenty-four hours. A lengthened cell cycle gives the genetic machinery of the cell more time to identify and repair genetic damage.

Although WR-2721 also gobbles up free radicals like other antioxidants, Grdina believes that enhancing the genetic surveillance and repair capacities of a cell is the primary reason why WR-2721 is effective in preventing genetic mutations. The biochemical properties believed to be responsible for the benefits ascribed to this compound involve its unique size, shape, and electrical charge—a combination of molecular properties not found in the antioxidants being promoted by advocates of prolongevity. Protecting against genetic damage in this way is vastly different and likely to be considerably more beneficial than just scavenging free radicals to reduce their number, the principal marketing claim used by some to sell vitamin supplements.

Over the years, scientists who have studied the effects of compounds like WR-2721 on the longevity of laboratory animals could have announced to the press that they had discovered an *elixir vitae*. Had they done so, reporters from around the world would probably have converged on their laboratories. Although the reporters would be intrigued by the military origin of these compounds, their attention would quickly switch to the small vial in their hands, and they would probably muse about what life would be like if everyone could remain healthy for more than a century. The investigators could have created a biotechnology company, and within days the stock price of their company would have soared as investors poured money into the discovery of a miraculous antiaging compound. None of this happened, and for very good reasons.

Investigators do not yet know if a compound like WR-2721 will inhibit the aging process in people. The animal and cell culture studies are extremely promising, but it is premature to make bold claims for their effects on humans. Although Grdina is aware that his research has implications for aging, his primary focus is on the development of treatments to lessen the risk of causing cancer in healthy cells when radiation therapy is used to treat cancer cells.

Although scientists who study compounds with antioxidant properties are cautiously optimistic about their possible use as a general antiaging compound in healthy people, most are waiting for clinical trials to be completed before making any such pronouncement. Scientists within both academia and industry are actively searching for compounds that minimize the accumulation of genetic damage within the cells of the body. This is real science, science about the biochemistry of life that will eventually have important applications for aging and longevity. In the ongoing search for antiaging compounds, the free radical scavengers now being sold and advertised as perfect protectors of DNA are unlikely to be significant players in the battle with aging, despite the boastful and exaggerated claims made by their proponents.

If compounds like WR-2721 are proven to have antiaging benefits, serious technical problems would still have to be surmounted in order to make these substances commercially viable for the general public. For example, WR-2721 must be injected. This is not a problem for patients already undergoing medical treatment, but the public wants an *elixir vitae* that can be swallowed in liquid or pill form. Injections of growth hormone have not become popular for this very reason, despite the exaggerated benefits claimed by their proponents. It is a difficult task to develop compounds that can get through the intestinal tract and into the bloodstream and still be able to enter through the cell membranes with the chemical properties necessary to react with DNA and perform their desired function. In addition, at present we don't know which is a more important factor influencing aging and longevity, genetic damage to the DNA within the nucleus of a cell, or damage to the DNA that resides within the mitochondria. If pharmaceuticals do end up being the modern equivalent of a Fountain of Youth, then compounds like WR-2721 that enhance the ancient mechanisms of cellular maintenance and repair could prove to be the *elixir vitae* mankind has sought for thousands of years.

THE FINAL FRONTIER

When the scientific dreamers of the twentieth century decided to make the fantastic idea of space travel a reality, they had to overcome many serious obstacles. The easiest among them was engineering and building a craft capable of reaching space. For manned space flight, the engineers also had to create a space capsule that would maintain a life-sustaining environment for the astronauts, protect them from the heightened radiation hazards in space, and return them to Earth safely. Although confident about the strictly mechanical challenges, the scientists at NASA were genuinely concerned that people might not be able to function or even survive in space.

A number of biological issues worried the NASA scientists responsible for the health and safety of the astronauts. One was whether the human eye would work properly in the weightless environment of space. This concern was heightened when John Glenn reported that he saw what looked like thousands of glowing fireflies outside the capsule during his first flight. NASA scientists also worried about the effects of weightlessness on other organs of the body. Would the heart, kidneys, lungs, digestive system, and so on function properly in space? At zero gravity, could a person do something as simple as swallow food and water? These concerns were all laid to rest when John Glenn returned safely from his first mission. From that historic point forward, it was known with confidence that human beings could function and survive in space, at least for a short time.

New health concerns arose as American and Russian astronauts began spending longer periods of time outside the gravitational force of the earth. It was discovered that two critical parts of the body undergo accelerated deterioration under conditions of prolonged weightlessness: muscle tissue and bones. The loss of bone mass and muscle tissue that occurs during weightlessness is an accelerated version of the normal consequences of aging. Although some pundits criticized the rationales given for the return of Senator Glenn to space in October 1998, monitoring the physiological processes of aging in an environment where they are accelerated can be of great value. By observing how the human body responds to the loss of bone mass and muscle tissue in space, and then learning how it recovers once back on Earth, we may achieve insights into how we can counteract these degenerative processes.

In terms of a scientific experiment, not much can be done

with a sample size of one. On the other hand, the collection of data on accelerated aging in space had to start with someone, and Senator Glenn proved an ideal test subject. He was thoroughly examined and monitored before, during, and after his flight, and this information was then compared with the wealth of recorded data on his physiology as a young astronaut. There is an appealing symmetry to the notion that scientists working to minimize health and safety risks for a handful of astronauts who have escaped the bounds of Earth may soon help the rest of us who are destined to spend our entire lives with our feet on the ground.

Earth's atmosphere is an excellent but imperfect shield from the most dangerous sources of radiation that travel through space. Prolonged space flights expose astronauts to much higher and considerably more dangerous levels of radiation. Materials like lead are commonly used to shield people from radiation in occupational settings such as nuclear power plants or medical and dental offices. Unfortunately, these materials are simply too heavy to haul into space in the quantities needed to protect astronauts fully from exposure to radiation. The prospect of missions to Mars, orbiting space stations, and other prolonged missions has made radiation monitoring, measurement, protection, and risk assessment some of the most important elements of research pursued by NASA scientists. Just as was the case with the military in the mid-twentieth century, NASA is exploring ways to develop better radio-protectors, as well as protocols for their safe use on astronauts. As with the study of "accelerated aging" caused by weightlessness, ongoing research by NASA will definitely expand scientific knowledge and may eventually lead to the development of potent pharmaceuticals that are relevant to aging-related diseases. If a genuine *elixir vitae* is to be found, it will likely be revealed as the product of practical necessity serving as the mother of invention.

ALTERNATIVE MEDICINE

It is important to distinguish between the genuine antiaging properties of products developed by scientists, and similar claims being made for products sold under the umbrella of alternative medicine. Alternative medicine is often perceived as an approach to illness derived more from superstition than genuine science. This attitude is ironic because many treatments found in Western medicine originated during a time of ignorance about human biology, and were developed and used before their biochemical properties were understood. As a result, most treatments for illness were discovered and tested using the age-old method of trial and error. A few hundred years ago, medical practitioners had no idea that bacteria and viruses that are invisible to the naked eye were responsible for most of the sickness and death that plagued humanity. Until the twentieth century, there was only a rudimentary understanding that the genetic codes for life were contained on chromosomes within the cells of the body. It is only a matter of time before the knowledge gained about the human body and the biochemical basis of disease may reveal that many traditional and what are now thought of as alternative treatments may be legitimate pathways to health and longevity.

In recent years, there has been a pronounced movement in the United States and other developed nations to embrace the practices of alternative medicine. Dr. Andrew Weil from the University of Arizona, a leading proponent of this approach, supports the legitimacy of alternative medicine by emphasizing the importance of the natural healing and protective powers of the body in a way that is identical to that of evolutionary medicine. Western physicians have responded to this shift by including courses on alternative medicine in the curriculums of medical schools. Scholarly papers have begun to appear in prestigious medical journals that document the magnitude and potential significance of this movement.

Many of the products currently marketed under the label of alternative medicine as ways to delay or reverse aging are simply worthless. It is unfortunate that legitimate developments are being obscured and diminished by hucksters trying to make a fast buck. Scientific evidence supports the claim that a number of chemical compounds derived from plants, animals, and other living things can have beneficial effects on specific diseases and symptoms of illness. There is, however, no scientific support for the claim that any food or supplement can either stop or reverse the aging process.

The shelves of health-food stores and grocery stores are filled with pills, powders, and liquids—each with a label claiming that it treats this disease or that symptom. It is important to remember that herbal supplements and vitamins are part of a largely unregulated industry, with few if any safeguards to protect the consumer. Many of these substances are biologically active, but the biological activity may have little or nothing to do with the claims on the label. Furthermore, recent investigations by the Food and Drug Administration (FDA) have found that the products you buy may not even contain the ingredients listed on the label. Studies have also shown that there is great variation in the concentration and potency of the active ingredients in these bottles. Because most of these substances have not been tested on either laboratory animals or people, the recommended dosage levels required to achieve the desired effect while avoiding toxicity or the encouragement of cancers are not known. The active ingredients of these substances can also interfere or negatively interact with medications already prescribed by physicians. Enough adverse health consequences have been linked to these so-called health products that many experts insist they should fall under the regulatory authority of the FDA. Until herbal medicines are regulated, about the best that can be said is, let the buyer beware—not a comforting thought when dealing with matters of personal health.

THE CLAIMS MADE by advocates of prolongevity that vitamins and minerals are the *elixir vitae* are false or greatly exaggerated. Most vitamin supplements are either excreted in the urine or they accumulate in fatty tissue. This means that megadoses of water-soluble vitamins produce expensive urine and megadoses of fat-soluble vitamins can be toxic. Although vitamin supplements certainly have significant health benefits for some people, there is no evidence to support the claim that they can forestall or reverse aging. Throughout history, people have managed to survive to very old ages without taking vitamin supplements, and those who died young usually died from an infectious disease. Malnutrition may have been a common problem in the past, but the solution to this problem for most people is a balanced diet, not the ingestion of vitamin supplements.

When many of the medicines still in use today were first discovered, pharmaceuticals were called elixirs, potions, and poultices, and the medical practitioners were called shamans, alchemists, and herbalists. These early practitioners of medicine were feared and revered, and their remedies were discovered by trial and error. They used materials from the world that surrounded them: soils, minerals, plant and animal products. For the substances that provided relief from illness or aided recovery, the practitioners of early medicine offered explanations that were shrouded in mysticism. There were no scientific advisory boards and no FDA to protect the citizenry from quacks. Many patients died from either the illness or the treatment, but some were also cured.

The search for therapeutic compounds in plants and animals continues today. In fact, the potential value of these compounds has bolstered efforts to maintain global biodiversity and preserve the rain forests. Identifying the genes that produce these compounds and discovering their structure and function reveals why they have therapeutic effects. Trial and error as well as luck still play a role as modern research replaces mysticism.

It is beyond the scope of this book to discuss the thousands of pharmaceuticals already being used to treat the symptoms of aging and disease. Pharmaceuticals are powerful weapons in the treatment of disease, and their medical importance will only grow in coming years. It is even possible that the aging process itself may fall under the influence of an *elixir vitae* developed in the laboratory. If such an *elixir vitae* can be discovered, it will most likely arise from a combination of solving practical problems unrelated to aging, such as those associated with exposure to radiation, and from research where antiaging is the primary goal.

The quest for immortality has now moved from folklore and legend to a frenetic scientific search for the biochemical keys that will unlock the secrets of aging. We have explained why there are scientific reasons to expect that substances with genuine antiaging properties may exist or are in the process of development, and how these substances might work to slow the aging process. Some of these chemical compounds will probably be available during the lifetimes of most young people alive today. However, it is important to distinguish between these genuine versions of the *elixir vitae* and the products being sold today under grossly exaggerated claims. It is tempting to believe that the keys to a long and healthy life are available now, but a realistic examination of the science of aging reveals that this is simply not the case. The legitimate science of aging has already led to remarkable extensions of life for many people. We can expect the hard work of researchers and medical practitioners to add to that success in the future.

CHAPTER 7

——— ✕ ———

The Genetic Frontier

*Whenever science makes a discovery, the devil grabs it
while the angels are debating the best way to use it.*

ALAN VALENTINE

THIRTY-THREE-YEAR-OLD WOMAN is spending a
day at the mall. After several hours of shopping, she
goes to the food court where she sits next to a mother
with three young children, one of whom is suffering from a
sinus infection. When the sick child drops a toy, the woman
picks it up and gives it back to her. The next morning the
woman wakes with the fever and body aches that typically
accompany a nasty infection. She is also four weeks pregnant.
The woman recovers in seven days, has an uneventful pregnancy, and delivers a healthy full-term boy weighing in at six
pounds four ounces. His name is Ricky.

At seven years of age, Ricky is an active child who excels in
sports and is talented in math and spelling. During a routine
physical exam, it is discovered that Ricky grew less than one

inch and gained only a few pounds over the last year. His pediatrician decides that Ricky's growth should be monitored. After six months without significant growth, he is referred to a pediatric endocrinologist. A series of blood tests reveal that he is producing only a fraction of the growth hormone that would be normal for a boy his age. He has no disease and is not sick, but an MRI reveals that a malformed pituitary gland—the organ that produces growth hormone—is the likely culprit for his stunted growth. Interestingly, his mother's exposure to a virus while pregnant with Ricky is one possible explanation for his problem. Whatever the cause, at his current rate of growth, Ricky will be only five feet tall as an adult.

Decades ago, when growth hormone (GH) deficiency was first identified, the condition was treated by injections of GH collected from the pituitary glands of cadavers. Children receiving these injections grew rapidly. The treatments, however, were expensive and the demand for GH exceeded its availability. The treatments were also not risk-free; some of the children receiving GH injections contracted Creutzfeld-Jakob disease. This lethal brain disorder was eventually traced to the cadavers, even though those people may have shown no signs of the disease in life.

The tools of molecular biology have become much more advanced over the last twenty-five years. Molecular biologists can now insert genes into bacteria and harness their reproductive power to create biological protein factories. Using this technology, proteins like GH can be manufactured in large quantities in a laboratory without any of the risks associated with GH extracted from cadavers. Further, the proteins manufactured by bacteria and those produced by the human body are virtually identical.

Thanks to genetic engineering, Ricky has a projected adult height of five feet seven—a height that is probably close to his genetic potential. While growth hormone replacement therapy is not a life-saving advance, many life-threatening diseases such as diabetes and cystic fibrosis are being fought using the same technology. Understanding gene structure, function, and regulation, and learning how to modify and manipulate these genetic attributes in order to preserve and enhance health and extend life, will form the foundation of the next longevity revolution.

TINKERING *with* MOTHER NATURE

Historically, most attempts to extend human life have involved the medical treatment of disease symptoms and protecting people from environmental forces of mortality. Michael Rose departed from this tradition when he extended the lives of his fruit flies through a program of selective breeding that favored unknown genes within the fruit flies that enhanced their longevity. This approach worked, but it took time, and the longevity benefit was accomplished indirectly by manipulating attributes of reproduction.

Scientific progress has opened the door to a fundamentally different approach. Discovering the rules that govern life at the molecular level will allow humanity to exert direct control over specific genes for the first time in history. This technology has the potential to enhance health and extend longevity by allowing us to augment gene products that diminish with age; to suppress the action of harmful genes; to remove damaged or harmful genes and replace them with desirable ones; to amplify the action of genes that enhance health and longevity; and to predict which individuals are at risk for genetic diseases.

AMPLIFYING EXISTING LONGEVITY

Rose demonstrated that by modifying the rules for breeding, he could produce generations of flies that lived longer than each preceding generation. Of course, even the flies he started with could live longer than their relatives in the garbage dump because they were raised in a protected environment and given a nutritious diet. Although Rose created a genetically altered fruit fly, he did not create a totally new kind of fly. Instead, he successfully amplified a genetic potential to live longer that already existed within some of the flies. A similar potential for greater longevity must also exist within the human gene pool.

The genes (alleles is more accurate) that are *associated* with greater longevity in people are scattered throughout the gene pool—more in some individuals and fewer in others. Individuals with the right combinations of these alleles have the genetic potential to live longer. The Rose experiments theoretically suggest that the frequency of these alleles within the gene pools of future human populations could be increased by selective breeding. Interestingly, studies conducted by Tom Perls from Harvard and David Snowdon from the University of Minnesota suggest that women who reach menopause at older ages tend to live longer than women who experience menopause at younger ages. The current explanation for this phenomenon is that prior to menopause, high levels of the female sex hormones (estrogen and progesterone) provide some protection from the health risks associated with aging.

If the experiment conducted by Rose could be performed on people, reproduction would be restricted to increasingly older women in each subsequent generation who remain capable of producing children. The perfect mates for these women would be the oldest men, who, by virtue of their great longevity, have proven that they too possess the right combination of genes for long life. Unfortunately, the effects of age would also prevent many of these men from participating in the experiment. The next best thing would be to choose young males born to long-lived parents. Why young males? Because teenage males are more likely to have sperm that have not accumulated genetic damage. Although there is little doubt that this experiment in social engineering would lead to a rise in human life expectancy, it would take hundreds of years to see the effect.

A faster way to increase human longevity would be to have all conceptions take place in a test tube. At puberty, germ cells (eggs and sperm) would be collected from every individual and then frozen for long-term storage. People who died before the age of sexual maturity would make no contributions to the supply of germ cells. The germ cells used to create future generations would be restricted to those from people who lived to extreme ages—all other germ cells would be discarded. The obvious problem with this approach is that it would require surrogate mothers and more than a century before a decision could be made about which germ cells to use for the test tube fertilizations. This approach also threatens the genetic diversity that provides a necessary reservoir of biological responses to unpredictable and hostile environments.

Of course, these approaches to enhancing longevity would not only be unethical, they would be unachievable. Reproduction relegated to the test tube invokes images of science fiction movies and Nazi experiments in human breeding. The human population is too big and people's sexual practices are too promiscuous to achieve the control needed to make such an experiment in social engineering successful.

AUGMENTING "PROTECTOR GENES"

Denham Harman from the University of Nebraska, the father of the free radical hypothesis of aging, has demonstrated that mice live longer when fed a diet supplemented by substances thought to reduce the quantity of free radicals. However, until the biochemical details of the protective mechanisms of the body are better understood, it is unlikely that artificial substances can perfectly mimic the behavior and effectiveness of those produced by the body. Even if substances like those produced by the body could be developed, getting them to the site where the genetic damage occurs (inside the nucleus of the cell where the DNA is located) remains an unsolved technical problem. To be effective, the protective compounds in dietary supplements must either survive the digestive process of the stomach or be converted by digestion to an active form of the chemical that actually provides the protection. The compound must also have properties that permit it to be absorbed into the bloodstream so that it can be delivered to cells. Upon arriving at a cell, the compound must then be capable of passing through the outer membrane of the cell and the inner membrane that forms the nucleus of the cell. The size, shape, and electrical charge of the man-made compound may not permit passage through these membranes without altering one or more of the physical attributes that influence its therapeutic or protective effectiveness.

Biomedical researchers have already discovered one new approach to increasing longevity that, at least for fruit flies, is better than using dietary supplements. Fruit flies, human beings, and other species have a gene that produces a substance known as superoxide dismutase (SOD) which scavenges free radicals. The difference between SOD and antioxidants ingested in the form of dietary supplements is that SOD has been developed and proven effective in the laboratory of life.

Under the assumption that if a little is good, then more must be better, a duplicate copy of the gene that produces SOD was inserted into the DNA of fruit flies. The genetically altered flies produced more SOD, which led them to live longer—an outcome that provides further support for the free radical hypothesis of aging. Promising as this research sounds, this kind of genetic manipulation may not work in people. Some early scientific data has linked overproduction of SOD to forms of amyotrophic lateral sclerosis (ALS) in mice, motor neuron disease in humans, and toxic side effects. Because genes typically perform multiple functions, enhancing one function without adversely impacting some of the others is a difficult, although not impossible, scientific and technological challenge.

In the future, additional "protector" genes will undoubtedly be identified and their roles in the defense of the genetic machinery of the cell revealed. Although significant technical obstacles remain, augmenting "protector" genes may eventually prove to be a highly effective way to enhance the genetic potential for longevity that already exists within the human genome.

SINGLE-GENE DISEASES

The technology of genetic engineering has incredible potential for the treatment of diseases caused by only one or a few genes. Single genes are known to be responsible for a wide variety of inherited human diseases, including Duchenne muscular dystrophy, cystic fibrosis, neurofibromatosis, retinoblastoma, Tay-Sachs disease, phenylketonuria (PKU), Huntington's, and many others. Additional diseases linked to specific genes are constantly being identified. In the near future, hundreds and perhaps thousands of disease-causing genes will have been discovered by scientists now working on the Human Genome project. One of the primary goals of the biomedical revolution will be to control or eliminate diseases and disorders by modifying or eliminating the genes that cause them.

This is not the only path to health and longevity. For example, in his book *The Lives to Come*, Philip Kitcher predicts that direct tinkering with genes will be rare. He believes that genetic interventions will be much more devious than finding and directly repairing or replacing harmful genes.

Eliminating the harmful genes that have been lurking within the human gene pool for millennia is not a realistic goal at this time. Fortunately, eliminating the health and mortality consequences of a disease does not necessarily require eliminating its underlying genetic cause. Most diseases progress through a complex series of biological changes and pathologic consequences as they proceed from initial onset to full manifestation. The bad news is that the complexity of disease processes often makes them difficult to understand. The good news is that this same complexity offers numerous opportunities for intervention. By interfering with one or more of the disease pathways, it should be possible to disrupt or neutralize a disease—a therapeutic approach to genetic manipulation that blurs the distinction between treatment and cure.

Lactose intolerance (the inability to digest dairy products) arises when the gene that normally produces the enzyme that breaks down milk sugar ceases to function. This genetic condition can be easily controlled either by manufacturing milk with the enzyme added or by taking a tablet containing the enzyme before consuming a dairy product. In the case of lactose intolerance, or the story about growth hormone that began this chapter, the genetic condition can be treated by acquiring the missing gene product from a source outside the body.

Scientists are already creating what are termed *transgenic animals*—animals that have been modified to carry a specified gene in every one of their cells. Like bacteria, transgenic animals can be used to produce therapeutic proteins. For example, transgenic pigs have already been created that produce human fibrinogen, an important component needed for the clotting of blood. A combination of transgenic animals and cloning creates opportunities for medical interventions that are not possible with genetically altered bacteria or viruses. Pigs have an immune system similar to humans and their internal organs are also comparable in size. It may soon be possible to clone transgenic pigs in order to produce a nearly unlimited supply of organs for transplantation into humans, organs that would not be rejected by the human immune system because they would carry the genetic material of the intended recipient.

Diseases and disease symptoms can also arise because too much of a gene product is produced. For example, overproduction of a gene product that helps nerve impulses travel from one nerve to another is one factor contributing to Alzheimer's disease. The excessive excitation of nerves in the brains of people suffering from Alzheimer's makes coherent thought almost impossible. Although we haven't yet found a cure, new drugs currently on the market can delay the progression of Alzheimer's either by binding and interfering with the gene product itself or by blocking the action of the gene that produces it. As the details of how gene malfunctions disrupt normal biochemical pathways become better known, the ability to influence the diseases they cause will improve rapidly.

Some therapies that treat rather than cure genetic diseases will have an effect on the transmission of genes to the next generation. Consider someone born with a lethal childhood disease like cystic fibrosis (CF). Every child who develops CF carries two copies of the harmful gene—one received from each parent. Historically, children with CF died before they were old enough to produce children of their own. Now, modified viruses can be used to insert a normal gene into the somatic cells of children with CF. This therapy extends their lives beyond childhood without eliminating the CF gene that resides within their eggs or sperm (germ cells). Therefore, any child they produce will receive a copy of the CF gene, leading to an increase in the frequency of the CF gene in future generations. Do not misconstrue the previous statement. One of the prime directives of biomedical research is to discover treatments for all genetic diseases, regardless of who carries them or the age at which they strike their victims. It is, however, important to understand that gene therapies will not always eliminate or reduce the frequency of harmful genes in the gene pool, and some will inevitably lead to a rise in the frequency of harmful genes, making the diseases they cause more common in future generations. This by-product of genetic manipulation is one of many ethical concerns that will have to be confronted as progress is made in the battle with inherited diseases.

ACQUIRED GENETIC DISEASES

Most people who hear the words "genetic disease" think of diseases that are caused by genes passed from parent to child, such as Down's syndrome, cystic fibrosis, Huntington's disease, sickle cell anemia, and Tay-Sachs disease. When a patient fills out a questionnaire at the doctor's office, propensities for diseases that have a heritable component are also being sought out—such as heart disease and cancer that run in a family. There are thousands of these so-called Mendelian diseases and disorders, named for Gregor Mendel, the Austrian monk who published the first theory of heredity based on the existence of genes in 1866. Many more genetic diseases lurking within the human gene pool are certainly waiting to be identified by the Human Genome Project.

Some people are surprised that there are so many heritable diseases. In reality, these diseases are relatively rare compared with the bewildering number of other ways people can become ill or die. So many, in fact, that disorders and deaths caused by heritable diseases are rare by comparison. However, the fact that heritable diseases are rare does not mean that genetic diseases are rare. We have already referred to the free radical hypothesis of aging—a theory which links diseases of aging to genetic damage that is accumulated over the course of a lifetime. Although this view of aging-related diseases is still described as theoretical, an expanding mass of scientific evidence collected over a forty-year period has convinced most scientists that the theory is valid. At present, it is impossible to say exactly how many deaths result from diseases caused by accumulated genetic damage because they are still being identified; but there is little doubt of their growing significance in an aging population.

Acquired genetic diseases arise from unrepaired genetic damage that accumulates over the course of a lifetime. As people experience increasingly longer lives, this damage has more time to accumulate and the number of people afflicted with these diseases will increase. Genetic damage that accumulates over time poses several significant challenges to the techniques of genetic manipulation currently being developed, as well as to those envisioned for the future.

Cancer and cardiovascular disease account for about seven of every ten deaths in developed countries. Some of these diseases almost certainly arise from the accumulation of genetic damage during the course of life. For example, humans and other animals possess a gene that has the critical job of causing some cells to commit suicide—a process known as programmed cell death or apoptosis. Apoptosis is an efficient way for the body to eliminate cells that, for a variety of reasons, are either no longer needed or no longer function normally. For example, during gestation, the human fetus has webbed hands and feet, but the cells that cause the webbing are instructed to commit suicide early in the developmental process. This cell suicide gene can be damaged over time by free radicals. When damage occurs, the gene loses its ability to instruct cells to commit suicide. The result can be uncontrolled cell growth, one of the defining characteristics of cancer.

Scientists and doctors are uncovering an increasing number of links between disease and the disruption of gene function or gene regulation. This raises the exciting prospect of preventing disease by directly repairing or replacing genes that are damaged during the course of life. One approach involves rendering viruses harmless so that their reproductive machinery can be used to deliver and insert packages of "repaired" DNA to damaged cells. At present, it is not possible to direct a modified virus to a specific cell or cell type, nor is it possible to insert the gene or genes into a specific location on a specific chromosome. If these obstacles can be overcome, this technology could yield near-miraculous cures for numerous diseases.

For some, the discovery of the importance of acquired

genetic diseases is good news because it follows that reducing the quantity of free radicals would lead to reductions in disease. Advocates of extreme prolongevity rely heavily on this relationship to promote substances that they contend will scavenge free radicals, eliminate disease, and thereby greatly increase life expectancy. Realistically, supplements of antioxidant vitamins may reduce the risk of some diseases by postponing them to later ages for some people, but the chance that any single individual will benefit is negligible. The life expectancy of populations would certainly not be influenced by vitamin supplements. Acquired genetic diseases result from genes that are inherently vulnerable to damage, environments that are inherently damaging, and bodies that constantly produce damaging substances as an unavoidable consequence of operating the machinery of life. Ultimately, the unique combination of genes and environment determines how long individuals live, and the harnessing of the body's own mechanisms to protect and repair vulnerable genes holds the greatest promise for achieving the next quantum leap in the pursuit of healthier and longer lives.

CHANGING *the* RECIPE *for* LIFE

In organisms that reproduce sexually, all life begins with a fertilized egg. From this humble microscopic beginning, the initial cell grows through a series of cell divisions into a wondrously complex organism comprised of many specialized cells. Any harmful gene that is present in the initial cell is thus passed on to every cell of the organism. In addition to these inherited genetic effects, each cell also accumulates its own unique and randomly located genetic damage. It is difficult to imagine how any biomedical technology can improve upon the incredible capacities that cells already possess to find and repair genetic damage acquired over the course of a lifetime. It is also difficult to imagine how any technology can address genetic damage occurring independently and randomly within trillions of cells. Clearly, the two best opportunities for genetic intervention would be in the germ cells prior to fertilization and in fertilized eggs before they start dividing. Of these, manipulating the germ cells prior to fertilization poses the fewest complications.

Although extremely controversial, the technology already exists to remove "unwanted or harmful" genes within eggs and sperm and replace them with "desirable" genes. The advantage of this approach is that every cell in the body would possess the desired genes. The disadvantage is that artificial fertilization (known as in vitro fertilization, IVF) must be used in order to ensure that the modified egg or sperm participates in the fertilization. IVF technology will probably never be used on a broad enough scale to have a significant impact on the human gene pool. However, for infertile couples and couples worried about heritable diseases, IVF is feasible and can be enormously beneficial. Unfortunately, many potential beneficiaries are currently denied access to this technology because of its high cost, an ethical concern with any medical treatment.

Scientists believe that it may soon be possible to replace or repair genes contained within germ cells using the same virus delivery system that has already been used to repair genetic damage within somatic cells. Using viruses to deliver packets of genes to a germ cell would give new meaning to surrogate parenthood. Couples could produce children whose genes do not come from either parent—an unprecedented biological phenomenon that could revolutionize reproduction and certainly alter the scientific meaning of evolution.

Although manipulating the germ line promises to reduce disease, improve health, and increase longevity, this technology is a Pandora's box of the worst kind. Ethical concerns abound. Probably the greatest misuse of this technology would be manipulating genes other than those that cause disease. Creating designer babies with preselected physical and physiological traits will be among the first of Pandora's evils. Choosing gender, eye, hair, and skin color, or attributes associated with intelligence, height, and athletic ability, may eventually become as easy as picking out wallpaper for the nursery.

The technology will certainly be used to manipulate genes associated with disease, aging, and longevity. If changes are made to the germ line, future generations will possess genetic attributes that were selected by people instead of nature— genetic changes made without their permission. Even though the technology of genetic engineering is in its infancy, the capacity to manipulate the recipe for life is already a controversial issue. It will become even more controversial when the pool of identified genes (and their functions) becomes larger, and methods of automation make the technologies of genetic manipulation simpler to use and more widely available.

The WISDOM of NATURE

It is difficult to pronounce judgment on just how wise nature has been with its filtering system, which permits a wide array of genes into each new generation, including those that kill children and people in the prime of their adult lives. To be accurate, nature does not actually move or shift the gene pool in any direction with a particular goal in mind. Individual members of every species and the genes they carry "succeed" if they and their offspring can survive and reproduce in the environment within which they live. This indifferent approach used by nature to propagate genes across time has led most species that have ever existed to go extinct.

Human beings have developed the technology to usurp nature's control over our genetic legacy, but this technology could do more harm than good. For example, it may become possible to eliminate genes that are linked to the frailty, disability, and disease associated with aging. If these genes also participate in critical developmental pathways early in life, then their elimination would have catastrophic consequences. It is uncertain whether deliberate decisions made by people about what genes should be passed on to future generations will lead to a "purified" gene pool and a disease-free population, or to catastrophic consequences for humanity.

IN *WONDERFUL LIFE* (1990), Stephen Jay Gould wrote that if the tape of life on earth could be replayed, the planet would be populated by a vastly different spectrum of organisms. Unpredictable historical accidents—events that Gould calls "contingencies"—would change the history of which species would persist and which would go extinct. Nature is a passive filter; neither cruelty nor benevolence directs the course of evolution. We have now gained the knowledge that will permit us to circumvent the process which has determined the success or failure of all previous genetic experiments in the laboratory of life. It does not matter how the newly discovered ability to manipulate the genetic code book is measured against ethical standards that exist today or in the future. There is no doubt that this technology will be used. If used before clinical trials reveal its full consequences, it will likely do more harm than good.

As scientists continue to discover more human genes, they will study these thoroughly in order to reveal not only their role in causing disease but also their potential role in promoting health and enhancing longevity. Eventually, these efforts may make it possible for every individual to have a personalized genetic profile stored in an accessible global database, or even placed on a microchip implanted in an inconspicuous location on the body. The profile would identify genes linked to heritable diseases, as well as any genetic propensities for diseases such as cancer and heart disease. Prior to a health crisis, this information could be used to suggest lifestyle choices that could minimize specific disease risks for that person. During a medical crisis, the information could help physicians diagnose and treat the cause of the crisis. As the technology of genetic manipulation improves and costs decline, it is hoped that everyone will be able to share in the benefits of this revolution.

It took millions of years of evolution for us to become what we are today. The incremental genetic changes that have occurred along the way have proven advantageous through eons of testing in the laboratory of life. Despite this, there will be incredible pressure to use the technology of genetic engineering to extend life, regardless of the potential negative consequences for individuals or society. The confrontations between emerging medical technologies and human ethics will no doubt escalate as a result of the battle against aging, disease, and death.

Circumventing or accelerating the rules of nature for the purpose of genetic change may prove to be the greatest accomplishment of our species. On the other hand, the history of human interactions with nature seems to consist mainly of repeated attempts to recover from the consequences of human error. For now, this technology has been used primarily to convert other organisms into biological factories that produce therapeutic agents needed by mankind. Genetic engineering is already one of the most sought after technologies of science because of its potential to show immediate, dramatic benefits for human health. Conversely, gene manipulation will be one of the most contentious and controversial issues ever confronted by human society. Like any other technology that has the potential for great benefit, genetic manipulation requires constant vigilance to ensure that it is not used either intentionally or inadvertently to cause great harm.

CHAPTER 8

——✂——

Long Life, Fleeting Youth

*I can do anything now at age 90 that I could do when
I was 18. Which shows you how pathetic I was at 18.*

GEORGE BURNS

T HERE IS A PROPHECY in Judeo-Christian tradition
that death will eventually be abolished: "He will swallow up death forever, and the Lord God will wipe away
tears from all faces" (Isaiah 25:8). According to scripture,
immortality would lead mankind to abolish war, embrace universal peace, and realize the full potential of mankind. If
swords are to be beaten into plowshares—another passage
from Isaiah (2:4)—it will occur because death will be defeated,
not because of a calculated plan to bring forth peace. The religious view of immortality is nothing short of spectacular. Why
would immortality induce mankind to abolish evil from the
world and embrace universal peace? Because once the heavy
burden of aging and death is lifted from the psyche of
mankind, the only barriers to enjoying the fruits of everlast-

ing life would be those that we create for ourselves. In a world where immortality reigns, the pursuit of any course of action that causes harm to ourselves or others would be foolish.

Our species has unparalleled foresight, but one of the liabilities of being able to contemplate and plan for the future is knowing that our own personal battle with death will be lost some day. Survival is a primal instinct that appears to be deeply ingrained within all living things. Almost without exception, death is fought to the very last breath. Recognizing that life must end, people have developed philosophies and religions to prepare mentally for death. Fields of science and medical institutions have been created that are dedicated to delaying death. Mankind has always fought the diseases that end life and will never stop seeking ways to delay death.

The biomedical revolution that will modify both aging and disease has already begun, and it will proceed at an accelerated pace in the twenty-first century whether or not anyone is prepared for it. Although this revolution will bring countless benefits, it also carries risks that must be weighed as we expand and accelerate efforts to prolong life.

The IMMORTALS *of* LAPŪTA: STRIKING *a* BARGAIN *with the* DEVIL

In 1726, Jonathan Swift published *Gulliver's Travels*, a series of delightful satires on politicians, philosophers, and scientists. The stories chronicle the remarkable adventures of Gulliver as he travels to exotic and remote nations. In one of these stories, Gulliver travels to Lapūta (pronounced La-pū-sha), where he encounters an unusual group of people known as Luggnaggians. On rare occasions, a Luggnaggian child is born with a red dot on its forehead, a signal that the child will be immortal. These Immortals, also known as Struldbruggs, fascinated Gulliver.

Gulliver fantasized about how wondrous life must be for people who learned early in life that death was no longer a threat. Living without the anxiety of aging and the fear of death must surely lift a great psychological weight from their minds, he thought. Immortality would allow them to pursue wealth and knowledge for centuries on end, a freedom that Gulliver believed would make the Struldbruggs the most revered oracles of their society. Although Gulliver "cried out as in a rapture" over the wondrous implications of immortality, the Luggnaggians considered him an imbecile for failing to recognize the true nature of everlasting life. For those who experienced it, immortality was a curse. Gulliver made the mistake of assuming that the Struldbruggs would forever be "in the prime of youth, attended with prosperity and health." In reality, their curse was to "pass a perpetual life under all the usual disadvantages which old age brings along with it."

The Immortals led a normal life until about thirty years of age, and then "grew melancholy and dejected." The emotional burden of knowing that the angel of death would never visit them caused the Struldbruggs to become extremely envious of mortals. By age eighty, they began to deteriorate rapidly. Their memories failed and their teeth and hair fell out. Eventually, they became total strangers in their own country because a loss of short-term memory prevented them from keeping up with the changes in language that were required to communicate with the younger generations. In his clever satirical style, Swift was mocking the philosophers of his day who argued that immortality, or at least a much longer life, was both possible and desirable.

Although Swift's story of immortality was intended as satire, his vision may have been prophetic. We have made unprecedented progress in extending life expectancy during the twentieth century. More people than at any other time in history are exploring the relatively uncharted older regions of the human lifespan. Are these explorers healthier than their counterparts in the past because of improved lifestyles and greater access to medical care? Or, like the Immortals of Laputa, has the modern rise in life expectancy exposed older generations to a minefield of diseases and disorders that were rarely experienced in the past?

The MEDICAL APPROACH: HURDLING ONE DISEASE *at a* TIME

The health of individuals as well as populations has always been influenced by the approach used to counter the forces that threaten human health. Today, good health and the survival of children are taken for granted in many countries. Recall, however, the grim historical fact that the vast majority of the billions of people ever born have died from an infectious or parasitic disease early in life.

It is a natural instinct to respond to an immediate threat. For medical practitioners throughout most of human history, that threat has been infectious disease. Confronting the most immediate health threat has fostered a largely reactive approach to disease. First the symptoms of disease appear, then they are treated. Historically, proactive measures have been either absent or only partially effective because the knowledge required for successful preventative measures simply did not exist prior to the eighteenth century.

The remarkable increases in life expectancy achieved during the twentieth century occurred only after proactive measures were adopted. These included the development of clean water, sewage disposal, refrigeration, controlled indoor living and working environments, and immunizations. The proactive approach has been an appropriate and incredibly successful response to the health threats posed by infectious and parasitic diseases. At least in developed regions of the world, these threats to early survival have now become overshadowed by cardiovascular disease and cancer, diseases that usually threaten the health and quality of life of people at older ages.

Now that great gains in longevity have been achieved, we have reverted back to a reactive approach to medicine. Most people seek the advice of a physician only when experiencing the symptoms of a disease. If the symptoms either subside on their own or are "cured," then the tendency is to return to a normal daily routine until confronted by the next health crisis. In other words, our personal response to aging-related health threats is just as reactive as the typical medical response. The cycle of symptoms and treatment continues until some insurmountable health crisis leads to death. Physicians and the public they serve have grown accustomed to this hurdle approach to disease. But the hurdle approach is just as inappropriate for the "new" aging-related threats to health that are common among long-lived populations as they were for infectious diseases that plagued earlier generations. The time has come to shift the attention of modern medicine to attacking the seeds of aging that are sown early in the lifespan of every person.

Early in the twentieth century, it was easy to predict one of the consequences of eliminating or reducing the threat posed by an infectious or parasitic disease: the lives of children would be extended by many years. This was a worthy and desirable goal for public health, and we all have benefited. Without the medical advances that made this goal attainable, most people alive today would probably have died early in life, just like most of our ancestors. Obviously, if children live longer, then they must eventually die at older ages. In other words, death has been redistributed from the young to the old. Given so little experience with the consequences of aging prior to the twentieth century, it was not so obvious what adults would die from. Scientists now know that the trade-off for saving the young from infectious diseases is a dramatic rise in chronic degenerative diseases among the elderly. Even with the benefit of hindsight, most people would consider the health consequences of the present to be an acceptable trade-off for reducing or eliminating those experienced in the past. Still, we should consider the loss of independence that many suffer in old age as we chart our course into the future.

What would the consequences for health be if heart disease, cancer, and stroke could be eradicated? Would the additional survival time manufactured by technology and improved lifestyles be healthy years, or would this additional time be dominated by increased frailty and disability? Would the health consequences in the absence of heart disease, cancer, and stroke be preferable to those that we experience now? In the zero-sum world of aging and death, a life saved today inevitably leads to another disease and cause of death tomorrow. The potential answers to these questions should be discussed even though technological progress will proceed regardless of the answers. History teaches us that society was unprepared for the medical and financial consequences of individual and population aging brought forth by the First Longevity Revolution. The cautionary tale of the immortal Struldbruggs spun so brilliantly by Jonathan Swift in the eighteenth century may be more prophetic than anyone can imagine because a long life has already proven to be a curse when not accompanied by good health.

HEALTH TRENDS: GETTING BETTER *or* GETTING WORSE?

Gerontologists are aware of the health issues associated with living longer and are actively monitoring the health and mortality consequences of individual and population aging. Two opposing schools of thought have emerged. One group supports what is referred to as "the compression of morbidity hypothesis," a controversial concept developed by Dr. James Fries, a physician from Stanford University.

In 1980, Fries predicted that future advances in medicine and the adoption of healthy lifestyles would simultaneously reduce death rates from killer diseases and postpone the onset and progression of nonfatal disorders (morbidity). Nonfatal disorders include such factors as arthritis, the loss of vision and hearing, osteoporosis, incontinence, Parkinson's disease, dementia, and a host of other conditions that may not be lethal but certainly diminish the quality of life. Fries postulated that 85 years of age represents a fixed biological limit to human life expectancy. He predicted that deaths in the future will become progressively more concentrated within a narrow age range around his proposed limit, leading to a compression of mortality. He then ignited a controversy by predicting that improved lifestyles would also concentrate the frailty caused by nonfatal disorders into the same narrow age range as mortality, a compression of both mortality and morbidity. This optimistic vision of the future should sound familiar; it is nearly identical to that predicted by Luigi Cornaro in the sixteenth century.

Many researchers felt that the compression hypothesis was naive. Although the mathematics of life expectancy suggest practical constraints on how high it can rise, there are no demographic data to support the existence of an absolute limit to life expectancy for populations, and there is no biological justification for the existence of an absolute limit for the longevity of an individual. The imposition of an artificial limit to the lifespan predicted by Fries led him to underestimate the number of people who can survive past age eighty-five. This is a critical point because people who live more than eighty-five years are confronted by a veritable minefield of aging-related diseases and disorders that rise in frequency with increasing age. It is also a matter of record that the number of frail elderly people with nonfatal conditions has been on the rise in recent decades.

The critics of the compression hypothesis emphasize that without the development of more effective treatments for the nonfatal diseases of aging, continued reductions in death rates from fatal diseases will increase the number of people experiencing a degraded quality of life in old age. These opposing sentiments were captured by a school of thought based on "the expansion of morbidity hypothesis," originally developed by Ernest Gruenberg and others in the 1970s, and later expanded by numerous other scientists. The adherents of this school also believe that death rates from killer diseases can continue to be reduced by changes in lifestyle and medical technology. However, they argue that the age at onset and the progression of nonfatal diseases will not be significantly influenced by reductions in the death rates for killer diseases. They also reject the notion of a fixed biological limit to life.

Removing the longevity barrier that concentrates mortality in the "compression" hypothesis opens the door to what has become referred to as an "expansion of mortality." This theory suggests that as life expectancy rises in the future, variation in the age at death will become greater because a more genetically diverse population will be surviving into older ages. As people benefit from reduced death rates, they not only live longer, they may also remain older longer. This extension of life gives the nonfatal disorders more time to further degrade the quality of life, leading to an "expansion of morbidity." Just because death from a killer disease is delayed does not mean that the underlying disease process has been similarly delayed. It may simply mean that medical technology is keeping a person alive longer with disease. Medical technology could be transforming fatal diseases into nonfatal chronic diseases, particularly at the end of life when the risk of frailty and disability is at its highest. Lois Verbrugge, a gerontologist from the University of Michigan, aptly summarized the message of the "expansion" school with the expression "trading off longer life for worsening health."

These two competing hypotheses cannot both be correct. The implications for human health are so important that researchers around the world have been gathering data and developing methods of analysis to discriminate between them. An international organization of researchers dedicated to this effort was formed in the 1980s under the direction of a French demographer, Jean-Marie Robine, funded by universities and government agencies in Australia, Canada, England, France, Italy, Japan, the Netherlands, Switzerland, and the United States. The term *health expectancy* was coined to refer to a division of life expectancy into the estimated number of years that are spent either healthy or unhealthy. During the 1980s, researchers examining global trends in health expectancy discovered that longer life was accompanied by worsening health, as predicted by the expansion school. However, by the 1990s, the evidence began to tilt back toward the compression school—a small improvement in the health status of the elderly was detected. Some researchers used this short-term improvement as a reason to declare that a new and favorable trend in the health of the elderly had not only emerged, but would continue into the future. The debate between the two schools of thought is not over and it is premature for either school to declare that they know what the future holds in store for us.

A New Paradigm: Successful Aging

Gerontologists are sensitive to the negative consequences that a bleak message about aging can have on how people think, both young and old. In 1987, Dr. John Rowe of the Mount Sinai School of Medicine in New York and Robert Kahn of the department of psychology at the University of Michigan popularized the concept of "successful aging." Although disagreeing on the details, the gerontology community began to appreciate the optimism generated by the idea that aging is not just a process of deterioration, and that aging-related causes of frailty and disease are more amenable to modification than generally believed. Subsequently, the prestigious MacArthur Foundation in the United States invested $10 million for research projects devoted to the positive aspects of aging.

The focus of research on successful aging is the identification of the characteristics shared by people who survive to older ages with their mental and physical health intact. These characteristics were defined as the ability to perform common activities of daily living (ADLs) such as walking, toileting, and dressing. Researchers want to discover why some people can continue to perform ADLs into extreme old age (successful aging) while others, by comparison, grow frail and disabled over time. It is their hope that ways can be found to make "successful aging" available to everyone—a laudable goal given that the elderly population will mushroom as baby boomers reach older ages early in the twenty-first century. Rowe and Kahn have published an excellent book entitled *Successful Aging*, which describes the concept. Their efforts to identify the underlying biological mechanisms responsible for successful aging have just begun.

As with any broad concept, successful aging has its limitations. The term is defined by the physial and mental attributes possessed by a subgroup of the population that the investigators identify as having aged successfully. People who do not age in this way have, by definition, not aged successfully. Not everybody can be John Glenn; he is an extraordinary person who has performed extraordinary deeds, along with countless others who have lived healthy and productive lives well past the age of seventy. Most young basketball players emulate Michael Jordan, but it is unlikely that any of them will match his skills and success. Defining success by the attributes of a highly select subgroup dooms most people to "failure," which can be damaging even in the absence of attached blame.

The successful aging concept has been heavily influenced by studies which suggest that the health of older people improved in the United States and some other developed nations during the 1980s. This is good news, and hopefully an ongoing trend, but it is risky to predict future trends from a selected subgroup of the population observed for a brief period of time. People who are eighty-five years and older today were born at a time that preceded many of the important benefits of the First Longevity Revolution. They survived two world wars, the Great Depression, several major influenza epidemics, polio, diphtheria, lower standards on food quality, poorer nutrition, high rates of cigarette smoking, and countless other hazards. In other words, today's eighty-five-year-olds represent a highly selected subgroup of hearty survivors. People living beyond age eighty-five in the future will not be as highly selected because many of the perils faced by today's elderly will have been either eliminated or dramatically reduced. In addition, a significant number of people in the future will live to old age only because medical technology will manufacture survival time for them, technology that did not exist for generations that aged successfully in the past.

The promoters of successful aging suggest that the optimist vision of aging seems to hold true and that this optimistic vision will apply to future generations as well. Anyone in their right mind hopes that they are right; but hope and reality do not always coincide. It is simply too early to declare that all future generations are going to age successfully. A conclusion on this important issue will have to await observation over the next three decades of an elderly population that will be far more numerous and diverse than at any other time in human history.

Hard Times *for* Entitlement Programs

In 1965, a machine was built that performed the remarkable task of extracting waste products from the bloodstream of people whose own kidneys were no longer able to perform this function. The treatment, commonly referred to as *dialysis*, was a lifesaver for those suffering from end-stage renal disease (ESRD). When the first dialysis machines were built, about 15,000–20,000 people of all ages were diagnosed with ESRD in the United States. Until then, the only treatment was a kidney transplant. Donor kidneys were rare and the biomedical knowledge about organ transplants was in its infancy. This meant that the vast majority of people with ESRD were destined to die within approximately six weeks of diagnosis.

The dialysis machine was a miraculous lifesaving technology, often adding decades to life; but this new technology also created a dilemma. Dialysis machines were dreadfully expensive and not enough could be built to meet the demand. As a consequence, most of the people who needed a dialysis machine did not have access to one, and many of those who did have access could not afford the cost of treatment. The dialysis machine was one of the first technological advances to trigger what has become an all-too-familiar conflict between the availability of lifesaving technologies and the ability to pay for them. The perplexing ethical dilemmas spawned by biomedical progress will only grow in number and complexity as societies grapple with continued individual and population aging.

In the early 1970s, the trustees of the Medicare program decided to make dialysis available to everyone in the United States who needed it, regardless of a person's age or their ability to pay for it. They accomplished this compassionate goal by using funds from Medicare to pay for the mass production of dialysis machines and cover the costs of using them in the clinical treatment of ESRD. Unfortunately, the projected size of the future beneficiary population needing dialysis was grossly underestimated by the analysts at Medicare. Due to unanticipated population aging, Medicare is now covering the cost of various treatments for over 275,000 people diagnosed with ESRD. In the two decades since dialysis became a reimbursable medical expense, actual costs have exceeded anticipated costs by billions of dollars, and the increase in expenditures is growing at a far faster rate than predicted just a few years ago.

The Social Security program is suffering a similar fate. In 1935, the creators of Social Security predicted that no more than 25 million people would ever draw benefits from the program. By the middle of the twenty-first century, population aging will cause the number of beneficiaries to exceed the original forecasts by at least 50 million. Debates over the future solvency of Social Security have thoroughly captured national attention. Underlying these debates is a demographic assumption that life expectancy in the United States will not exceed 85 years by the year 2080. Given the constraints imposed on mortality by entropy in the life table, this assumption about life expectancy seems reasonable. However, if future advances in the biomedical sciences lead to increases in life expectancy at a pace that is even a fraction of that achieved during the twentieth century, then life expectancy could rise beyond 85 years. If this happens, age-based entitlement programs will require even more dramatic restructuring.

CIRCUMVENTING *the* RULES *of* NATURE

In an earlier chapter, we discussed two ways in which the biology of living things can be altered: by natural or artificial selection over extended periods of time or immediately by the direct manipulation of genes. Through artificial selection, Michael Rose was able to demonstrate that he could produce generations of fruit flies that lived longer than each preceding generation. However, when the long-lived flies developed in the laboratory were returned to the garbage dump, they died out more quickly than their outdoor relatives because they were unable to withstand the rigors of living outside the laboratory. The flies had sacrificed hardiness for longevity, an exceedingly high price to pay.

For centuries, we have been striving to create environmental conditions for ourselves that resemble a laboratory. Progressively more effective barriers are being erected between human beings and the world in which we live.

Controlling nature is an ambitious goal, and the dramatic gains in life expectancy of the last two hundred years are a remarkable testament to the progress that we have made.

The biomedical technologies of the future will generate changes in health, disease, and death in a fundamentally different way. Scientists are learning to influence health and longevity directly by suppressing the action of harmful genes, compensating for the effects of abnormal genes, and even replacing harmful genes with more desirable ones. These are not subtle genetic changes occurring within populations over long periods of time, but changes within individuals that will have immediate and potentially dramatic health benefits. The age of genetic manipulation has already begun.

The technology of genetic engineering has great potential to influence both health and longevity by treating disease, especially heritable diseases caused by only one or a few genes. Once the allelic form of a gene that causes a heritable disease like cystic fibrosis is known, a variant of the gene that does not cause disease can be inserted into the genetic machinery of a virus that has been rendered harmless. The virus can then be used to do what viruses do best—insert their genes (now including the desired human gene) into the DNA of the afflicted individual. Because the genetic changes made by the virus are restricted to somatic cells, they cannot be transmitted to offspring and their effect ends with the life of the individual. Nevertheless, some scientists have raised concerns over the potential danger of mixing human and viral genes. The sudden death in 1999 of one of the patients receiving this experimental treatment has heightened such concerns. Others have expressed fears over what might happen if this technology is used by rogue scientists who either do not consider or ignore the potential health risks. At present, a consensus within the scientific community on the potential genetic risks associated with this technology has not emerged. As a result, scientists are proceeding with extreme caution, as are the government agencies that fund their research.

Another way to use genetic engineering for the purpose of combating disease and extending longevity is to manipulate the germ line itself—the DNA contained within eggs and sperm. Again, this approach involves removing unwanted genes and replacing them with genes judged to be "desirable." Since every cell in the body is derived from a single fertilized egg, this approach has the advantage of giving every cell in the body the desired replacement gene.

The greatest concerns with a technology that can modify the germ line are ethical. Although costs may drop in the future, will they drop enough to include all potential beneficiaries, regardless of socioeconomic status? Will the technology be used to manipulate genes other than those that cause disease? For example, how can safeguards be established to ensure that genetic manipulation will not be used to create designer babies? The capacity to manipulate the recipe for life is already a controversial issue even though the technology is not fully developed.

Circumventing the rules of nature in order to make rapid genetic change may prove to be the greatest technological accomplishment of our species. But tinkering with nature also opens up a Pandora's box of biological and ethical concerns of such importance that they demand extreme caution and prudence.

KNOWLEDGE AND GENES are propagated in fundamentally different ways. Every individual of every generation has received genes that were present within the gene pool of the previous generation. Knowledge, on the other hand, expands and accumulates, allowing every generation to possesses knowledge that did not exist in the previous generation. The danger of rapidly expanding knowledge is that the power it creates can exceed the ability to use it wisely. Mankind has acquired a new and awesome power, the ability to manipulate the basic building blocks of life. This technology will advance rapidly during the twenty-first century, and as is typical of the capabilities spawned by knowledge, the technologies of gene manipulation are going to be used whether or not the public understands them, or society is prepared for the consequences of their use.

We could have succumbed to sensationalism and devoted this entire chapter to such topics as the possibility of a human-constructed virus escaping into the general population, or the use of genetic engineering to create a new human species. Instead, we chose to focus upon the immediate consequences and challenges created by the next longevity revolution—scientific progress in the biomedical sciences that is making it possible to manipulate our biological destiny.

The First Longevity Revolution has already permitted an unprecedented number of people to enter into the relatively unexplored older regions of the lifespan. These survival gains arose primarily from a progressive control over environmental sources of mortality, mortality that has little to do with our basic biology. A consequence of being allowed to grow older has been a transition to causes of death that arise from flaws, breakdowns, and failures of the biological processes within our bodies. It will be nearly impossible to control the threats posed by these aging-related diseases with the traditional medical focus on the treatment of disease symptoms.

One of the primary goals of research in the biomedical sciences is to treat disease by curing it. Understanding the molecular basis of disease and devising methods of intervention at the level of genes is an important element of the next longevity revolution. But pursuing cures poses significant scientific, technical, and ethical challenges. It is inevitable that forthcoming biomedical technologies will be used in an attempt to extend longevity. The challenge will be to avoid what happened to the long-lived fruit flies in the experiments conducted by Michael Rose—increased longevity achieved at the expense of health and hardiness.

Serious social and economic challenges can also be expected to emerge. For example, the average retirement age in developed nations has been declining in recent years. If life expectancy continues to rise, people will be living an unprecedented one-third or more of their lives in retirement. This poses a daunting financial challenge for retirees, as well as for age-entitlement programs that were not originally designed to pay beneficiaries for decades. Marriage and divorce could also be profoundly influenced by greater longevity. For couples who retain their health, extended survival creates many opportunities for enhanced and extended physical and emotional pleasures.

The dark side of this scenario involves the people who do not survive to older ages with their health intact. For them, extended survival may not be a blessing and will place an extremely difficult emotional and financial burden on their spouses and loved ones. Women are likely to suffer the most because they tend to live an average of five years longer than men. In the twenty-first century, the next longevity revolution will force humanity to face the awesome burden of responsibility that comes with having gained an ability to manipulate the biological recipe for life. Perhaps the elder members of our society are the only ones who can bring wisdom and guidance to a world about to chart its own biological destiny.

CHAPTER 9

—— ⋈ ——

Longevity for Sale

The trouble with the world is not that people know too little, but that they know so many things that ain't so.

MARK TWAIN

THE POPULAR LITERATURE about aging and longevity touts the miraculous healing powers and life-enhancing properties of a large variety of foods, waters, vitamins, minerals, hormones, chemicals, and spirituality that are available to all of us today. These are the same sorts of magical substances and ideas that appeared within the Fountain, Antediluvian, and Hyperborean legends common throughout the history of the prolongevity movement. Although this movement has made important contributions to public health, there have always been and there will always be those who exploit longevity for profit. From the unscrupulous alchemists who concocted exotic elixirs to the hucksters of the nineteenth century who sold potions from the back of a wagon to the media-savvy pseudoscientists and spiritualists of today, there is no subject more amenable to the allure of magic and mysticism than the pursuit of longevity.

An extremely important difference between the pro-longevity movement of the past and the one that exists today is that for the first time in history, the human population is aging rapidly. With billions of people about to survive to older ages, the sheer number who could benefit from a genuine antiaging drug has sparked a frantic search for ways to fore-stall aging and disease. As a result, there is now an international obsession with the elusive Fountain of Youth. Fear of aging and death is a powerful emotion that can overwhelm the reason of even the most astute person, especially when offered a supposed miracle treatment for age or disease.

Feeding the prolongevity frenzy is a mass media hungry for the one great story that has eluded scientists for thousands of years. Imagine the attention and money that awaits the first person or pharmaceutical company that claims legitimately to have discovered a true Fountain of Youth. A moment like this occurred in 1998 when scientists discovered an enzyme, known as telomerase, with the potential to immortalize cells that would otherwise have a limited lifespan. Within days, news reporters were speculating that aging would soon become a thing of the past. The stock of the company support-ing the telomerase research soared. The fact that the scientists who made this important discovery never claimed it would influence the aging process, and that telomerase prob-ably has little or nothing to do with the aging and death of people, was not mentioned during the brief moment this research captured the attention of the media.

An immense scientific effort has gone into the legitimate study of aging. Scientists have already made and continue to make exciting discoveries about the inner workings of the aging process. Their research has led to interventions that permit most people today to live much longer and healthier lives than they would have in the past. Generally, before research findings are presented to the scientific community and the public, they have been carefully scrutinized and thoroughly discussed by knowledgeable experts. However, when the conservative and densely constructed prose of science is translated for public consumption, it can be easily misunderstood and is often distorted. Because scientific information is often veiled in statistics, it is difficult to discriminate between fact and fiction—recall "The Gambler's Odds" in chapter 3. One of our goals in writing this book is to help you make informed judgments about these claims, and then decide for yourself whether you need to change your lifestyle, or purchase and use the latest fad presented as a "scientifically proven" antiaging substance.

Let's focus on a few of the themes that are most popular today.

TODAY'S ANTEDILUVIAN LEGENDS

One of the most appealing of the Antediluvian approaches to prolongevity on the market today is the paleolithic diet. The legitimate science behind this idea was developed by S. Boyd Eaton and his colleagues from Emory University. Scientists are confident that humankind existed in its current form some 130,000 years ago. Eaton suggests that these early humans had diets containing vegetables, fruits, nuts, and berries, and large quantities of meat that, like modern range animals (such as bison or deer), would have been naturally low in fat. According to Eaton, if our early ancestors were well adapted to their paleolithic environment, then the human body of today must also be well adapted to a paleolithic diet.

Eaton contends that our genetic endowment is not designed for the fast-paced lifestyles, crowded living conditions, and reckless dietary habits of the modern world.

Archeological sites and the dentition of fossils can only provide hints about the diet of our early ancestors. Although deaths due to acute trauma may be visible as fossil evidence, diagnoses of deaths caused by diseases that affect soft tissues and organs are based on conjecture because time destroys the evidence. We can, however, gather indirect evidence of the potential health benefits of a paleolithic diet because isolated populations following this way of life still exist in remote areas of the world today. Assuming that their diet and lifestyle are similar to those of our early ancestors, the link between diet and health can be investigated by studying these modern hunter-gatherers.

Eaton compared the fitness, diet, and disease prevalence of fifty-eight primitive societies with comparable data collected from industrialized nations. As a percentage of total caloric intake, the contemporary American diet contains twice the fat and one-third the protein of diets maintained by the primitive populations. Eaton also discovered a lower frequency of cardio-vascular diseases and cancers among hunter-gatherers than for people living in industrialized nations. Of even greater relevance to the issue of health benefits, Eaton found that the further the lifestyles and diet of a population departed from those of pure hunter-gatherers, the more closely their health and mortality resembled that of the industrialized world.

Advocates of the paleolithic diet believe that the major diseases observed in the industrialized world are caused by departures from the diet to which our early ancestors were presumably adapted. They argue that these diseases could be eliminated, health restored, and our lifespans significantly extended if everyone would adopt a paleolithic diet. There are remarkable parallels between the health continuum associated with departures from the paleolithic diet and the Antediluvian theories of Roger Bacon. Bacon believed that the human lifespan was declining over time because the adverse effects of decadent lifestyles accumulated from one generation to the next. Like the modern advocates of the paleolithic diet, Bacon thought this condition could be reversed if people would adhere to the lifestyles of previous generations.

The advocates of the paleolithic diet tell a compelling story. However, if the paleolithic diet is so conducive to health and longevity, why are the life expectancies of primitive societies often decades lower than those of industrialized nations? One reason is that without access to antibiotics, infectious and parasitic diseases continue to be the major killers in developing countries and primitive societies. In addition, minor health problems in the developed world such as an abscessed tooth, a blister, or appendicitis can be life-threatening events in a society where access to medical care is limited. However, even when these people escape death by infection, accident, or predator, they still fail to live as long as people in developed countries. Why is this so?

The answer can be found in the lessons of evolution introduced in previous chapters. Organisms are molded by natural selection to survive long enough to reproduce. Long life in the hostile world of our ancestors would have been an extravagant waste of precious physiological resources better spent producing babies. The incredibly varied dietary habits observed among primitive societies today (eating and surviving on everything from tarantulas to whale blubber) demonstrate that our species enjoys an amazingly adaptable physiology that permits us to extract nourishment from almost any food source. There is no reason to believe that adaptation to a paleolithic diet would make it more conducive to disease resistance and great longevity because the selective pressures of evolution effectively cease after reproduction. The diseases we associate with poor dietary habits, such as heart disease and cancer, are typically experienced at older ages, after reproduction. Even if our early ancestors lived to old age—generally they did not—these late-life diseases would have played no part in human evolution.

There may be some health benefits (weight loss and a lower risk of certain diseases) to be gained among some people alive today if they follow a paleolithic diet. However, these benefits represent by-products of adaptations for everyday survival, not evolutionary adaptations for longevity without disease in a hostile world. Departures from a paleolithic diet simply cannot be the primary reason why there are high death rates from heart disease and cancer in industrialized nations. It is far more likely that improved living conditions and access to sophisticated medical care permit people in the modern world to live long enough to die from diseases that are associated with growing older. Although it is certainly true that some foods, such as refined sugars and animal fat, contribute to disease, there is neither convincing evidence nor scientific logic to support the claim that adherence to a paleolithic diet provides a longevity benefit.

When Less Is More

A modern version of an ancient Taoist method for extending life, caloric restriction, has won some ardent advocates in the world of science. In the 1950s, scientists demonstrated that rodents live 40–50 percent longer when the caloric intake of their diet is restricted, a finding that has been subsequently confirmed for a variety of other organisms. Recently, Richard

Weindruch at the University of Wisconsin has found evidence to suggest that the longevity benefits of caloric restriction may also apply to primates. The implication of his research is that caloric restriction has the potential to extend the lives of human beings. Legitimate studies like these are cited by pro-longevity advocates to bolster their claim that aging, like any other "disease," can be treated and eliminated.

Laboratory rats are not the lean, mean sewer rats that scurry through alleys anxiously looking for their next meal. Males and females are kept in separate cages, and food supplies are never more than a few inches away, so laboratory rats have little else to do but sleep and eat. The animals on normal diets just lay around and get fat, while those on a caloric restricted diet lay around and get less fat. Neither group is being compared to rats that live in the wild, the human equivalent of people who adhere to a vigorous regimen of physical exercise. In human terms, caloric restriction studies are equivalent to comparing the longevity of obese couch potatoes with people who otherwise consume a normal diet. As a result, these studies do not demonstrate the longevity benefits of caloric restriction as much as they illustrate the detrimental consequences of an indulgent lifestyle.

A more serious obstacle posed by caloric restriction diets is how to implement them on a broad scale. Several years ago, I attended an international conference in Paris devoted to research on aging. After a long day of presentations on the longevity benefits of healthy lifestyles and dietary supplements, an organizer of the conference invited the audience to attend what promised to be a memorable meal at a fabulous French restaurant. Tongue-in-cheek, I suggested that it would be more appropriate to dine at a health-food store than to head off to a gastronomic binge. Everyone laughed. Later that evening, in traditional French style, we feasted for hours on goose liver pâté, duck, steak, rich pastries, and large quantities of wine. Clearly, it is far easier to advocate an abstemious (calorically restricted) lifestyle than it is to adhere to one.

Dr. Roy Walford is a well-known molecular biologist in the field of aging and the strongest proponent of caloric restriction. Walford practices what he preaches: he adheres to a caloric restriction diet and has stated that he anticipates living at least 120 years. In addition, he has published two popular books on the topic—*The 120-Year Diet* (presenting the legitimate science behind caloric restriction) and *The Anti-Aging Plan: Strategies and Recipes for Extending Your Healthy Years* (essentially a cookbook). Shortly after the cookbook was published in 1995, Walford hosted a widely publicized dinner party for the news media, publishing industry, and various paparazzi. His guests were served a broad sampling of his life-extending recipes. The general consensus of his guests was that the food was so awful that a lifetime spent adhering to the diet would seem like an eternity, even if it did not make you live any longer.

The value of research on caloric restriction is not going to come from encouraging people to restrict their intake of food. One benefit of this research has been to demonstrate that obesity and lethargy operate as potent accelerators of aging, with specific and often dramatic life-shortening effects. A caloric intake below the excessive levels that prevail in developed nations today may very well give rise to legitimate health and longevity benefits. However, future benefits to public health from this area of research will more likely come from identifying the underlying biological mechanisms that are responsible for the effect, rather than encouraging the adoption of diets that almost nobody wants to follow.

TODAY'S FOUNTAIN LEGENDS

Prolongevity claims based on Fountain legends constantly appear in daily newspapers, weekly tabloids, and in popular books that have sold in the millions. *Time* and *Newsweek* run lead stories on discoveries with a Fountain theme at least twice a year. Go to a local bookstore and you will find dozens of books, all published within the past five years, in which the authors claim either to have discovered the Fountain of Youth themselves or to have extracted its essential properties from scientific studies.

Fountain legends follow two basic themes. One is the age-old claim that some vital substance lost during the course of life, if replenished to youthful levels, has the power to rejuvenate the body. The second involves vitamins, minerals, and other substances that are claimed to have only recently been proven by science to prevent disease, restore youth, and reverse the aging process. Most people are probably familiar with these claims: avoid heart disease with vitamin E, eliminate cancer with vitamin C, reverse the aging clock with growth hormone, rejuvenate the body with melatonin, testosterone, and vegetable juice cocktails, and combat disease with the powers of garlic, vinegar, selenium, and magnesium.

Although the print media may be the traditional source of such information, it is not the only source. Undocumented articles about the discovery of antiaging substances abound on the Internet, where they spawn virtual chat rooms filled with enthusiastic people and impassioned discussion. Antiaging medicine has even invaded the aisles and walkways of your local grocery stores and malls. Recently, I spotted a long line of people waiting at a booth to purchase a vitamin supplement whose label claimed it would stop the aging process and prolong life. Out of curiosity, I joined the queue. When my turn came, I asked the attractive young woman in the booth how much longer I could expect to live if I took the vitamin. Without hesitation, she said a good six to ten years, but only if I stuck to the required daily (and expensive) regimen for the rest of my life. For centuries, people have been sold this same bill of goods as they gathered around the back of a wagon waiting their turn to buy some magical elixir.

Vitamins, Minerals, Supplements, and Other Antiaging Products

The basic argument made by those who claim that vitamin supplements have antiaging properties is that aging is an unnatural disease process that, like many other disease processes, can be slowed down and even reversed. Aging is portrayed as little more than a deficiency disease that can be treated. This line of reasoning has a familiar ring to it. It is a combination of Luigi Cornero's philosophy that mankind is healthy by nature, Taoist beliefs in the presence of *hsien* (immortalizing) substances, and Frances Bacon's view that ingesting potions would restore youth by reversing the aging process.

Scientific arguments have always been difficult to understand, but they are particularly complicated in the biological realm of aging, which is based on changes in molecular processes that cannot be observed by the human eye. Scientific explanations have also become more statistical in nature, and as we've said before, medical statistics are easy to misinterpret. This is not to say that scientific information can be properly interpreted only by scientists. Many science reporters have written excellent books and newspaper articles that present complex scientific information in a way that is accurate and understandable to a broad audience. Unfortunately, advocates of vitamin supplements often extend the legitimate science upon which their views are based well beyond the views held by the scientists who discovered them. Not surprisingly, some authors who promote this form of longevity have produced their own multinutrient formulas containing the essential vitamins, minerals, and other substances described in their books.

The premise behind using vitamin supplements to forestall the aging process is that aging is caused by the accumulation of damage to DNA created by free radicals. Since vitamins and a variety of other compounds have been shown to neutralize free radicals, those who promote vitamin supplements as antiaging compounds claim that ingesting large quantities of these substances will stop or reverse aging. The logic is simple enough; the problem is that it will not work. In order to understand why, we need a more detailed explanation of the link between free radicals and aging.

The Free Radical Hypothesis of Aging

All living things require energy to stay alive. To get this energy, food must be converted into a form of energy that the cells of the body can use. The digestive system breaks food down to molecules small enough to enter cells. Once in a cell, the food molecules are transformed into useful energy by a series of biochemical reactions (metabolism) that takes place within microscopic structures called mitochondria. Unfortunately, the metabolic reactions within the mitochondria not only produce the desired energy; they also produce highly reactive and potentially damaging molecules known as free radicals.

Like iron filings drawn to a magnet, the magnetic charge of oxygen-derived free radicals causes them to be attracted to DNA. Two kinds of DNA exist in most cells—the DNA in the nucleus that contains the recipe for life; and a second set contained within the mitochondria. Both are targets for the destructive action of free radical molecules, and there is a growing consensus that both sources of genetic damage contribute to the aging process.

The good news is that free radical molecules are ephemeral, existing for just millionths of a second. The bad news is that they are inevitably and continuously produced within every living cell. When free radicals interact with nuclear DNA, they have the capacity to cause a mutation, a small change in the structure of the DNA molecule. Since free radicals are not guided missiles, the mutations they cause occur at random locations. The randomness of this genetic damage explains, in part, why people with ideal lifestyles can develop cancer early in life, while other people with poor lifestyles can sometimes live to extreme old age.

Free radicals are not the sole source of mutation, but their inseparable link to metabolism ensures that they are a constant threat. Scientists estimate that the DNA in every cell of your body is exposed to approximately ten thousand potentially damaging hits each day. Humans and other organisms are able to survive this assault because we posses remarkably efficient biological mechanisms that find and repair over 99 percent of the genetic damage that occurs. The lack of absolute perfection provides the scientific basis for the free radical hypothesis of aging, but also guarantees that the scavenging properties of vitamin supplements cannot significantly influence human longevity.

Scientists have demonstrated that some damaged segments of DNA are either improperly repaired or not repaired at all. According to the free radical hypothesis, it is the accumulation of genetic damage over time that inevitably leads to the failure of cells, tissues, organs, organ systems, and ultimately the individual. These unrepaired errors also contribute to the physical signs of aging that you see in the mirror.

The cells of our body have a first line of defense against the damaging effects of free radicals that is identical to the presumed action of the antioxidant vitamin supplements being sold today. Scientists have discovered genes that produce molecules referred to as "free radical scavengers." Just as white blood cells patrol the bloodstream in search of the unrecognized proteins that make up bacteria and viruses, scavenger molecules circulating passively within the cells are prepared to capture and disable free radicals before they can do harm. Scientific evidence is mounting that free radical scavengers may be the junior partner in the defense against free radicals. The senior partner is likely to be the molecular mechanisms designed for genetic maintenance and repair. Together, they form a formidable partnership that provides a nearly perfect defense against the threats posed by free radicals.

The proponents of vitamin supplements claim that their elixirs will provide a perfect defense against genetic damage by preventing the endless barrage of free radical missiles from hitting their target. The "Star Wars" debates during the presidency of Ronald Reagan demonstrated the mathematical folly of a "perfect" missile defense in a world brimming with nuclear weapons. According to a growing consensus within the scientific community, any improvements that can be made to the near perfection of the body's ability to defend against genetic damage are likely to be achieved by enhancing the existing maintenance and repair processes of the cell—not by ingesting antioxidants, or anything else that is intended only to protect DNA before damage occurs. Two things are clear: It is extremely difficult to improve on near perfection; and without free radicals, life as we know it could not exist. Even if vitamin supplements could neutralize some or most of the free radicals present in the body, those remaining would continue to lead to the accumulated damage that contributes to aging.

There is likely to be a biological reason why nature did not develop perfect protection from mutations within the DNA of eggs and sperm, including those caused by free radicals. These same mutations are partially responsible for evolutionary change and species adaptation. Most mutations are harmful, and selection operates to reduce their frequency in the population. However, once in a while a mutation offers a survival advantage in a given environment, and people who carry this gene have a reproductive advantage. In other words, mutations create the genetic diversity needed to survive and thrive in a changing environment.

There is also a reason why it may not be a good idea to dramatically reduce free radicals, even if vitamin supplements could be made to do so. Free radicals are known to participate in such important biological processes as the immune response, cell proliferation, intercellular communication, and cell death. The complete elimination of free radicals would disrupt the normal functioning of the body rather than extend it. In the end, the biological machinery that inevitably generates the energy needed to sustain life just as inevitably sows the seeds of its own destruction.

Can Everyone Live to Be 120?

Advocates of extreme prolongevity have claimed that the vitamin and mineral supplements they sell will permit everyone to achieve 120 years of age, the maximum human lifespan perpetuated by folklore. Luigi Cornaro also held the mistaken belief that everyone is born with the same lifespan potential. No credible scientist today believes this to be true. Every lottery ticket has a unique number, and every person has a unique genetic potential for longevity. Even identical twins develop genetic differences as the DNA of each twin accumulates its own unique mutations. Although someone will eventually live past the current record of 122 years, the odds that you possess the right combination of genes, environment, and luck to be that person are even longer than winning the Powerball lottery.

It is worth remembering that for thousands of years some people have lived to extreme old age under much harsher environmental conditions than exist today, without any of the supplements now being promoted by the prolongevity advocates. It is an irresistible temptation for authors to exaggerate and for publishers to hype in order to capture the attention of readers in the highly competitive world of bookselling. Nothing else would seem to explain the claims that aging can be reversed and that everyone can enjoy a lifespan of 120 years. The truth is that while we have increased the number of people living to old age, there has been only a marginal increase in the maximum age to which the healthiest and luckiest people live.

Hormones: The Fountain of Youth or the Ultimate Hype?

The fatalists and early prolongevists believed that aging occurred because of the loss of some vital substance. The Taoists called this substance *ching* and "vital breath"; Hippocrates, Aristotle, Galen, and Avicenna referred to it as the "innate moisture of the body"; Bacon called it "essence"; and for Cornaro it was an unspecified substance known as "vital principle." All of these descriptions probably arose from observations of the visible signs of aging—drying of the skin, shrinking of the skeleton, loss of muscle mass, hair loss, and decline in sexual activity.

Methods designed to either conserve, replenish, or replace a vital substance lost during the course of life are the defining characteristic of remedies derived from the Fountain legends. In the past, advocates of prolongevity concocted elixirs and potions containing secret substances they claimed would maintain or restore youth. Today, the elixirs and potions have been replaced by hormones such as growth hormone (GH), melatonin, DHEA, pregnenolone, testosterone, estrogen, and progesterone. Promoters claim that restoring these hormones to their youthful levels will not only stop the process of aging but actually reverse it.

For those who promote the anti-aging properties of hormones, aging is once again defined as a disease that is amenable to treatment, but free radical damage is replaced by declines in the level of one or more hormones as the main cause of aging. Those who promote the longevity benefits of GH view aging as a deficiency disease of the pituitary gland; they postulate that injections of GH will eliminate this deficiency, restore youth, and extend the functional human lifespan to somewhere between 120 and 130 years.

The first important link between GH and aging appeared in an article written by the endocrinologist Dr. Daniel Rudman in the *New England Journal of Medicine* (1990). Dr. Rudman injected twelve men between the ages of sixty-one and eighty with GH three times a week for six months. When compared to men of comparable age who did not receive the treatment, those receiving GH experienced a gain in muscle mass, a loss of fat, improved skin elasticity, improved sleep, lowered cholesterol, and a number of other beneficial changes that made it seem as though the Fountain of Youth had finally been discovered. Needless to say, this study received enormous international attention.

The beneficial effects of GH injections were not a fluke. Studies demonstrating the health benefits of hormone therapy have been conducted by responsible scientists for years. The problem is not the science, it is how the science has been interpreted.

Rudman observed that the men receiving GH therapy in his study developed physical and physiological measures of aging that were comparable to individuals ten to twenty years their junior. Observations like this made Rudman and others believe they had overturned the prevailing dogma that the loss of physiological function was an irreversible aspect of growing older. These findings have been erroneously interpreted by the proponents of GH therapy to mean that the aging process had been *reversed* and that GH had literally made the test subjects younger.

The claim that we can all grow younger has been made by many prophets of prolongevity, not just those promoting GH. The unfortunate reality is that the biological clocks of aging move in only one direction: forward. There are no magical substances that can be ingested or injected, no behavioral practices that can be adopted, which will reverse aging, any more than time can be reversed by resetting the calendar on your watch back to an earlier date. The damage to DNA, cells, tissues, and organs caused by random hits of free radicals can only increase and accumulate during the course of life.

Those who claim that hormones and other substances can make people grow younger confuse the levels of physiological fitness that can be achieved at any age with resetting biological clocks to a younger age. The use of hormones, pharmaceuticals, and behavioral practices can make your body look like it did when you were younger; but looks can be deceiving. The

unfortunate reality is that beneath these visible improvements, damage continues to accumulate at the molecular level,
the processes responsible for maintaining the functional
integrity of the body persistently degrade, and a host of other
internal signs of wear and tear accompany the passage of time.

On what basis do the purveyors of various hormones make
their claims? Most rely on studies that use laboratory animals.
For example, scientists have demonstrated that laboratory
mice receiving the hormone DHEA live 40 percent longer on
average than mice that do not receive the hormone. Advocates
claim that people would receive the same longevity benefit. At
present, the highest life expectancy anywhere in the world is
about 80 years. Increase this number by 40 percent, and the
conclusion is reached that DHEA can raise the life expectancy
to 112 years, an increase of 32 years.

At first glance this conclusion may seem reasonable, but
several problems can arise in extrapolating results from laboratory animals to humans. One of the most serious is that the
laboratory animals used in scientific experiments are often
genetically identical whereas human populations are always
genetically diverse. While genetically identical animals may
respond to a treatment in a similar way, our genetic uniqueness ensures that we will not. From "The Gambler's Odds"
(chapter 3) we are reminded that some people will receive
dramatic benefits, some will be harmed, and most people will
fall between these extremes.

Comparisons between laboratory animals and humans are
further complicated by the kinds of environments in which they
live. Unlike people, laboratory animals are raised in highly controlled environments that minimize health risks. As a result, the
human response to any treatment will always be more variable
than that observed among laboratory animals. In the case of
DHEA, there is even reason to believe that the longevity benefit

may not have been a result of the hormone. Because the animals in these studies did not like the taste of DHEA, the longevity effect may have been a product of caloric restriction rather than the hormone itself. There is no question that animal experimentation plays a pivotal role in the development of medical interventions that help us and the identification of environmental hazards that cause us harm. However, extreme caution must be taken when using animal studies to predict the magnitude of the human response to treatments intended to influence phenomena as complex as aging and longevity.

Many proponents of hormone therapy also suggest that the broad use of antioxidants will add seven years to the thirty-year benefit claimed for hormone therapy. Unfortunately, they have made the mistake of adding together the longevity benefits reported for a variety of antioxidants and hormones. Health risks associated with cigarette smoking illustrate why this is problematic. Smoking increases the risk of dying from a number of diseases: emphysema, heart disease, stroke, and several forms of cancer. For each of these diseases, scientists have estimated how much death rates would decline if people stopped smoking. Unfortunately, when you look at real mortality statistics, the actual mortality reductions from all causes of death combined are considerably less than the sum of the reductions estimated for each smoking-related disease considered separately. If antioxidants and hormones acted upon different processes that affect longevity, and these processes themselves did not interact with each other, then their longevity benefits might be added together. Since this additivity is extremely unlikely, the longevity benefit claimed for a combined regimen of antioxidants and hormones would be exaggerated, even if based on reasonable estimates of the benefits of each.

Although many physical and physiological benefits have

been attributed to GH therapy, all of these benefits and more can be attained by most people free of charge with no shots, no blood tests, no costly office visits, and minimal side effects. A simple exercise program can increase muscle mass, reduce fatty tissue, improve sleep, slow down bone loss, improve mental acuity, and much more. Although the health benefits of exercise have been known for thousands of years, only recently have scientists such as Dr. Marie Fiatarone of Harvard demonstrated that many of these benefits can be achieved by men and women of any age, including the extreme elderly.

The physical and physiological benefits of hormone therapy are documented in the scientific literature. In their proper place, with careful supervision, hormone supplements can have a beneficial effect on the quality and length of life for some people. However, why anyone would bother with these shots or pills is hard to understand because exercise can provide even more benefits, and it costs little more than effort. Advocates of this form of therapy cannot justify their claim that hormones, even when supplemented by antioxidants, can cure, modify, reverse, or in any way significantly influence the human aging process. More importantly, there is no scientific evidence to support the claim that GH or any other *elixir vitae* will make anyone live longer than they would have anyway.

TODAY'S HYPERBOREAN LEGENDS

Whenever we give a presentation about aging, someone invariably asks: "But what about the people in Soviet Georgia who live to more than a hundred and twenty years by eating yogurt?" We find this humorous because a TV commercial for yogurt has added yet another substance to the long list of products that perpetuate Hyperborean legends.

Since Hyperborean legends have already been thoroughly refuted in both the scientific and popular literature, we will not go into detail on this topic. *How and Why We Age*, by Leonard Hayflick, and *Why We Age*, by Steven Austad, are excellent sources on the subject, as are the scientific articles published by demographers Neil Bennett and Lea Garson.

Today, common Hyperborean legends include stories of long-lived people from the Caucasus Mountains of Soviet Georgia, the Andean village of Vilcabamba in Ecuador, and the Hunza who live in Kashmir. The main problem in all of these cases is that there is no proof that anyone in these areas is the age that he or she claims. With no written language, no detailed records of births, marriages, or other relevant events bearing a reliable time stamp, it is impossible to document anybody's age. Combine this with societies that revere extreme old age, and local and nationalistic pride in longevity, and it becomes clear why and how these people exaggerate their claims. In one classic example, a scientist who returned to the village of Vilcabamba four years after his first visit found that the oldest person in the village claimed to be eleven years older than when first interviewed.

For years, the *Guinness Book of World Records* listed a Japanese man by the name of Mr. Uzumi as the oldest person in the world. He was said to have died at the biblical age of 120. His name has remained in the record book even though Mr. Uzumi may not have been who he claimed to be. In early twentieth-century Japan, on occasion a younger sibling would assume the complete identity, name and all, of an older sibling who died. In this case, the person claiming to be Mr. Uzumi may have been his younger brother, an impressive but not record-setting centenarian.

In the summer of 1997, a fascinating story about human aging and longevity made international headlines. The oldest woman in the world had died at the age of 122. Madame Jeanne Calment was born in southern France in 1875 and lived there her entire life. She smoked cigarettes for over a century and still rode a bicycle at 110. When her health began to fail, she moved into an assisted-care facility where, upon initial evaluation, she was diagnosed as having a form of dementia that made it difficult for her to communicate. It was later discovered that she only spoke and understood a rare dialect of French. When a physician who knew this dialect screamed a few words into her ear, the personality of a remarkably alert and intelligent woman emerged.

Although physically frail and deprived of much of her hearing, sight, and taste for the last twelve years of her life, Madame Calment was an exceptional case of human survival. What makes this story so remarkable is that her age has been scientifically verified. The French demographer Jean-Marie Robine, her physician Dr. Victor Lébre, and gerontologist Dr. Michel Allard spent countless hours poring through detailed and reliable historical records on Madame Calment, documenting her every move through life. Details of her story were published in the book *Jeanne Calment: From Van Gogh's Time to Ours* (1998). This is a remarkable achievement, to be sure, but it became a little less surprising when it was discovered that for generations, her ancestors were all extremely long-lived.

Almost without exception, every report of extreme longevity has either been tainted by doubt or has outright failed the test of scientific scrutiny. Because of the persistence of reports on extreme longevity, scientists have made repeated visits to places where Hyperborean legends abound. Although the lifestyles of these people are often simple, there is great variation in their diets. Some eat small quantities of fruits and vegetables while others consume large quantities of meat and dairy products. Ironically, smoking and the consumption of alcohol are common among the extreme elderly. This just shows once again that our genetic legacy has a more important influence on our lifespan than many would like to believe.

A SPIRITUAL APPROACH *to* LONGEVITY

One branch of the prolongevity movement is particularly interesting because it does not involve the marketing or sale of antiaging substances. Instead, proponents market a spiritual message. For example, some spiritualists claim that disease is unnatural and aging a conditioned response that can be overcome. In other words, staying young and living a long life require merely the proper state of mind.

Many spiritualists based their philosophy on the idea that all of the cells of a newborn baby would function perfectly, without aging, if it were not for the toxic debris that accumulates within cells during the course of life. Maintain a lifestyle based on spirituality and moderation, they say, and everyone can avoid aging. Pediatricians will tell you that the cells of newborns addicted to crack or thalidomide are not free of toxic debris. Ask women who have experienced spontaneous abortions and the mothers who have given birth to children with Tay-Sachs disease, early onset diabetes, childhood leukemia, and hundreds of other genetic diseases, whether the cells of their children are perfect. This false line of reasoning perpetuates the disproven philosophy of Luigi Cornaro that every person is born with the same biological potential to live a long and healthy life. It is almost as if people are being blamed for experiencing the negative consequences of aging.

Long ago, the discovery of genes overturned the notion that we are all born with a *tabula rasa* or clean slate of health upon which we write during the course of life. According to Philip Kitcher in his book *The Lives to Come*, every person carries an

average of ten to twelve genes with mutations that are potentially lethal—mutations that will be passed on to their children.
In the real world of genetic diversity, free radicals, radiation,
metabolism, and accumulated damage to DNA, no child is born
anywhere in the world with cells that can avoid aging.

The spiritualists also suggest that the biological phenomenon of aging is a product of our inherited expectation that
the body must wear out over time—a belief rooted in the ethos
that we are all fated to suffer, grow old, and die. This implies
that aging is largely a psychological phenomenon that can be
avoided. However, if this view is correct, then why do animals
other than humans grow old and die? As far as we know, animals are not capable of contemplating their own mortality,
nor can they formulate beliefs concerning their fate. Is the
declining vigor observed in other species a result of their
"expectation" that aging will occur? Overwhelming scientific
evidence demonstrates that aging is nearly universal for all
living things. The regularity of aging across vastly different
life forms strongly suggests that aging follows predictable biological rules with physical consequences, a biology that gives
rise to spiritual and psychological challenges, not the reverse.

The spiritual approach to health and longevity has been
influenced by Antediluvian themes which permeate the
ancient mythologies of India, China, Japan, and to a lesser
extent the Christian West. In some of these legends, ancient
masters controlled the aging process by manipulating the "life
force"—referred to as *Prana* in Indian folklore. This concept is
remarkably similar to Hippocrates' notion of "vital force."
Using a handful of scientific studies that link mental attitude
to disease, modern spiritualists declare that the mind and
body are one. Gaining control over the mind opens the way to
controlling the body, which in turn permits the "aging disease" to be controlled.

In the case of Indian folklore, *Prana* was thought of as the subtlest form of biological energy contained within all physical and mental planes of existence. Indian sages believed that they had discovered ways to both conserve and acquire *Prana*. It could be acquired through various foods and waters, physical exercise, appropriate behaviors and emotions, and an emphasis on "breath" that is nearly identical to Taoist beliefs. Conversely, *Prana* could be depleted by bad behaviors and negative emotions. Similar themes about some substance lost during the course of life, but which could be replenished through proper spiritual guidance, permeates all of the spiritual approaches to longevity. While these spiritual exercises may reduce stress and bring other mental and physical benefits, the simple truth is that none of them will significantly extend an individual's life.

How Long Do Prolongevists Live?

One of the most glaring ironies of the long history of the prolongevity movement is the observation that despite their claims, advocates of prolongevity have themselves died. If a genuine Fountain of Youth had actually been discovered, the first ones to benefit should have been the discoverers.

Some of you may remember Jim Fixx from the 1980s, the long-distance runner who advocated the health benefits of running. He died of a massive heart attack at age fifty-two near the end of a routine run. Remember Linus Pauling, who ardently promoted the idea that massive doses of vitamin C would "prevent" both colds and cancer? He succumbed to cancer at age ninety-three. Jerome Rodale was the founder of Rodale Press and *Prevention* magazine. He worked tirelessly to promote the health benefits of organic farming and declared confidently that he would live to a hundred unless killed by accident. During an interview with Dick Cavett in 1971 on the health and longevity benefits of organic foods, the seventy-two-year-old Rodale dropped dead of a heart attack. Dr. Daniel Rudman, the first scientist to test the antiaging properties of growth hormone, died at age sixty-seven from a pulmonary embolism. Of course it is not possible to know with certainty what the early prolongevity figures died from, but it is safe to say that most probably died from one of the four top killers that prevail today: heart disease, cancer, stroke, or an infectious disease.

Several years ago, an antiaging organization sent us a letter claiming that aging and death could be overcome and immortality achieved simply by abandoning the belief that they were inevitable. In fact, its members believe that there are already immortals walking among us today. They took exception to one of our articles in which we suggested that, given the current state of medical technology, life expectancy at birth will not rise above 85 years. Sometime later, when the founder of this organization died, his followers claimed that he had lost his faith. It is difficult to refrain from declaring that the advocates of extreme prolongevity, whether they promote spirituality or various Antediluvian, Hyperborean, or Fountain themes, suffer from "aging denial"—a reckless refusal (common among teenagers) to acknowledge and accept one's own mortality.

MESSAGES OF CHANGE and improvement resonate in a society that has contributed significantly to many modern technological marvels. Scientists and great thinkers of every era have overestimated the benefits to be gained from the advances in technology made during their time. People in every era have believed that their science would reveal the secrets of the Fountain of Youth. Today is no exception. As history repeats itself, there is no doubt that our science will also appear crude and overly optimistic to historians looking back into the twentieth century.

Futurists and prolongevity advocates claim that it is only a matter of time before aging and all of its accompanying diseases, disorders, and infirmities will be a thing of the past. They maintain that every person is born to be healthy; disease is a product of modern civilization and decadent lifestyles; and aging is just another disease waiting to be conquered. Have the prolongevists of today, with their vitamin supplements, hormones, and spirituality, really discovered the Fountain of Youth? Both history and science suggest that they have simply rediscovered the same old false claims and misleading promises used throughout history to exploit the desire of people to find youth in a bottle.

Mixing pseudoscience with exaggerations and distortions of legitimate science is certainly not new. However, an enormous mass media that can instantaneously captivate a global audience with sensational sound bites and eye-catching headlines is new, and it lends remarkable power to those seeking to profit from the sale of longevity. Those people who claim that we can live decades longer by making dietary modifications, by ingesting vitamins, antioxidants, and hormones, or by meditating, appear to us to be nothing more than modern versions of the hucksters and sideshow salesmen of the past.

CHAPTER 10

—⋈—

A Prescription for the Twenty-first Century

Gather ye rosebuds while ye may,
Old Time is still a-flying:
And this same flower that smiles today
Tomorrow will be dying.

ROBERT HERRICK

T HE OLD TESTAMENT TELLS of a time when mankind experienced no sickness or old age. In a two-thousand-year-old interpretation of the Bible called the Midrash, Jewish scholars wrote that during the era of the great patriarchs, people grew to adulthood and then remained physically youthful throughout their lives. When death did finally arrive, it occurred suddenly and painlessly on a day predetermined by heavenly decree, the soul departing through the nose with a sneeze. The phrases "Bless you" and *Gesundheit* offered today when someone sneezes are linguistic artifacts from this biblical story, originally uttered to prevent the soul from leaving the body.

According to Genesis (24:1) and its interpretation in the Midrash, the origin of old age can be traced to the patriarch Abraham. Abraham's wife was taken captive by Avimelech, the Philistine king of Grar, and was found to be pregnant shortly after her return, although the Bible states that Sarah was not touched by Avimelech. In order to stop the growing rumors over paternity, God decided to make the child, Isaac, in the exact image of his true father, Abraham. According to the Midrash, Isaac grew up to be physically indistinguishable from his father because aging after adulthood did not exist. This created a problem for Abraham's people. They could not properly honor Abraham because they were unable to distinguish between father and son. God solved this problem by making Abraham appear physically different from his son, and then decided to make all parents look distinguished by giving them flowing white hair and a physical appearance that changed with time.

According to Genesis (48:1) and the Midrash, the patriarch Jacob was responsible for introducing sickness to mankind. Jacob wanted to talk to his sons about some of their negative personality traits, but to emphasize the importance of this discussion, he wanted to deliver it on a memorable occasion, just before his death. To accomplish this, Jacob needed a warning that his death was imminent. God answered the prayers of Jacob by making sickness the sign that death was forthcoming. Apparently, God decided to make this warning available to everyone, perhaps to provide an opportunity for everyone to make peace with their loved ones. While we were writing this book, we asked our friends Rabbi Avie Shapiro and Rabbi Yisrael Koval why Jacob did not request something a bit more benign, like a rash. Their answer was simple: Because sickness did not yet exist, Jacob would not have known what to ask for. Besides, he would never have been so presumptuous as to give advice to God.

For thousands of years, visions of a Golden Age when aging and sickness did not exist have been captured in Antediluvian myths and people have dreamed of returning to this glorious age. Today is no different. Advocates of extreme prolongevity are still trying to convince people that their remedies can take us all back to the time before Abraham, Isaac, and Jacob. Although their promises are unrealistic and their claims exaggerated, the vision of a world without sickness and old age is a useful reminder that improving the quality of life should be the primary goal of biomedical research and medical intervention.

MEASURING TIME

How old are you? You probably think of the answer in years. Aging is obviously linked to the passage of time, and researchers who are interested in human aging also use years to measure how much time has passed since birth. In this book, you have already read such remarks as life expectancy at birth is 80 years, and the maximum recorded age at death is 122 years. However, measuring time in this way can muddle our understanding of aging. To illustrate, imagine a large room filled with people who are all sixty years old. Some of these people will appear to be far younger than sixty, while others will appear far older. Aging is a biological phenomenon, and it should be clear from this simple example that the passage of biological time is not the same as the passage of chronological time. People age biologically at different rates because the internal clocks they inherit from their ancestors have different tempos, and people differ in the way they choose to live their lives.

A related problem is that aging is a continuous process. Using a large unit of time like a year obscures the unseen biological changes that occur not only on a day-to-day basis, but on a moment-to-moment basis. Using a day to measure time is manageable, and makes it possible to view the aging process from a different and more realistic perspective. In order to make the following material more meaningful, take a look at the life expectancy of people your age and sex, expressed in days, in the life table in the appendix.

Regardless of how much we exercise, modify our diets, or alter our lifestyles, 90 percent of everyone born in developed countries like the United States will live somewhere between 22,000 and 32,000 days. Men live just over 27,000 days on average and women live about 29,000 days on average. Of the remaining 10 percent, most will die before 22,000 days, and a handful of Methuselahs will be born with the genetic potential to live as long as 40,000 days. This is the range of lifespans observed under the best survival conditions ever experienced by humans. These achieved lifespans are the product of a body design molded by the hostile environment of our ancestors, restricted by environmental and intrinsic forces of mortality, and influenced by medical interventions.

Throughout this book, we have argued that biological and mathematical forces place upper limits on the rise in human life expectancy. *These limits do not mean that babies are born with a fixed allotment of days as was believed in the time of the patriarchs.* Instead, the lifespan achieved by an individual should be thought of as falling within a range of possible days of life that can be lived because of his or her genetic longevity potential and the favorable conditions provided by the First Longevity Revolution. In a genetically diverse population like our own, the babies born in a given year will have a wide range of realized lifespans. Because every person is a unique genetic entity, with life experiences and a personal lifespan potential that cannot be predicted, the *potential* longevity that can be attained by any specific individual cannot be known. However, as the life table in the appendix indicates, the *average* longevity anticipated for individuals of a given age can be estimated from the collective longevity of people living in a particular time and place.

Invisible changes are occurring constantly in our genes, cells, tissues, and organs—changes that we eventually observe in the mirror and that ultimately affect our health and longevity. On a more philosophical level, measuring time in days forces us to acknowledge that life is lived one day at a time. The Roman poet Lucretius observed that no matter how long life is lived, it is insignificant compared to the infinite amount of time that is spent dead. Using this logic, Lucretius argued that instead of a constant struggle against death, life should be a never-ending search for new ways to appreciate each day that is lived. Anyone who has lost a loved one can appreciate his point. During the final days of each of my parents' lives, I [B.A.C.] experienced a near-debilitating fear of what might happen the next day. Although not learned soon enough, their deaths taught me an important lesson: Fear of tomorrow can cause you to miss moments of happiness that can be shared today.

ONE MORE DAY *of* LIFE

Every year, the National Academy of Sciences sponsors a meeting designed to bring together scientists from the biological, social, and earth sciences. The spirited discussion and interdisciplinary exchange that takes place at these extraordinary meetings is science at its best. At one of these meetings, a young demographer stood up and asked a question that went something like this: If a man lived for 32,873 days (90 years), could he have done anything differently during the course of his life that would have allowed him to live just one more day? His answer to this question was a resounding yes. Perhaps he could have eaten one less ice cream cone, exercised a few more hours, or taken a larger daily dose of vitamin E. If one day could be added to a life of 32,873 days, then it should also be possible to add one more day to a life of 32,874 days, and so on.

The young demographer extended his line of reasoning to make several additional conclusions. He suggested that almost all diseases and causes of death can be modified or eliminated. He also claimed that there is no scientific justification for the existence of a biological limit to the length of life of an individual. He went on to argue that even if such limits exist for individuals, there is no demographic evidence that limits on life expectancy for populations are being approached anywhere in the world today—not even in countries like Sweden and Japan that have the highest life expectancies. Ultimately, he suggested, even if limits to life expectancy exist, they are so far ahead of us that they are of no relevance today.

Initially, the "one more day" argument seems reasonable, but within the demography community, there is serious disagreement over whether there are upper limits to human life expectancy. Most researchers who reject upper limits base their opposition on predictions from mathematical models that give no indication that such limits exist. Researchers who argue in favor of upper limits on life expectancy point out that a population is comprised of biological entities with body designs that impose limits on their physical performance (speed, agility, quickness, and strength). If limits are a universal rule for attributes associated with body design, then an attribute like longevity must also have limitations. This leads to the inescapable conclusion that there are limits on the life expectancy of human populations regardless of the predictions made from purely mathematical models.

How many "one more days" can be achieved by further progress in controlling environmental forces of mortality, by additional survival time manufactured through biomedical technologies that not only treat but also cure disease, and by days of life gained through the identification and widespread adoption of healthy lifestyles? The ongoing debate over limits to life expectancy is not likely to be resolved any time soon. We still have much to learn about the forces of human health and mortality.

There is little doubt that reductions in deaths caused by hostile environments will continue to help more people achieve their lifespan potential. How far individuals will be able to surpass their lifespan potential will depend on the degree to which the biomedical sciences can manufacture survival time by overcoming the genetic legacy of disease and aging inherited from our ancestors. Within the biological constraints that make it impossible to operate living machines much beyond their warranty period, much can still be achieved by the medical miraclemakers.

Exercise and diet can improve both health and quality of life, but scientific studies have demonstrated repeatedly that life expectancy is not extended much for populations that have adopted "healthy" lifestyles—a benefit of approximately 900 days relative to those who have poor lifestyle habits. How can people who adhere to what are defined today as the healthiest lifestyles add more days to their lives? They already exercise vigorously every day, sleep eight hours each night, avoid tobacco products and the excessive consumption of alcohol, and they eat low-fat foods that include lots of fruits, vegetables, and fiber. Where is "one more day" going to come from for them? Some of these people will strive for that "one more day" by turning to the products being sold by the advocates of prolongevity—megadoses of antioxidant vitamins, blue-green algae, growth hormone, melatonin, and dozens of others. Unfortunately, there is no scientific evidence to suggest that these products will have any effect on human longevity.

ONE MORE DAY *of* HEALTH

Instead of asking whether there is anything you can do to live one more day, it is more reasonable to ask yourself whether there is anything you can do to be healthier tomorrow than you are today? Do not, however, confuse becoming *healthier* with growing *younger*. This is a common mistake made by some of the more ardent advocates of prolongevity. Growing younger would not only involve a vast array of significant physiological improvements; it would also require impossible reductions in the unavoidable health risks associated with the passage of biological time. However, most people of any age can choose to improve their health and preserve it for long periods of time. Imagine, for example, that you resolve to start a year-long fitness program that combines regular exercise with a nutritionally balanced diet. By adhering to this program, you would lose body fat, increase muscle mass, decrease the rate of bone loss, improve mental acuity, lower cholesterol levels, and receive numerous other health benefits. If this sounds a lot like the claims made by those extolling the virtues of spending thousands of dollars every year on daily shots of growth hormone, you are right.

If you can be healthier a year from now, can you be healthier tomorrow than you are today? The answer to this question is an emphatic yes! Exercise and a healthy diet offer benefits that are incremental and cumulative. A little exercise and a degree of dietary moderation every day is better than no exercise and a poor diet. Just as with the "one more day" argument, there are some advocates who contend that if you can be healthier tomorrow and even healthier the next day, then the gains in health made by improvements in lifestyle can continue indefinitely.

Unfortunately, the biology of aging and health is not that simple. An eighty-year-old man may, with considerable hard work and dedication, reclaim a level of physical and physiological fitness that would have been far easier to attain when he was sixty years old. However, no matter how hard he works, a lifetime of wear and tear on his joints will never allow him to reclaim the fitness potential that was available to him when he was twenty. A seventy-year-old woman may achieve a strengthening of muscles that would be the envy of a fifty-year-old woman, but she can never achieve results comparable to those that would have been available to her had she made the same effort when she was fifty years old. Although a fitness program adhered to throughout life can neither negate nor reverse the aging process, it can help slow down the progressive loss of attainable fitness levels that accompany advancing age.

BIOLOGICAL LOOPHOLES

A recurring theme throughout this book has been that nature does not "build" the genetic programs that make up different kinds of organisms. "Building" implies an active intent whereas the forces of nature act like a series of passive filters that genetic experiments in life, like people, must pass through in order to propagate their genes. As such, nature cannot build a genetic program for either aging or death. Old age and sickness are accidents, not phenomena that occur by conscious design. A theologian might view these phenomena as answers to the prayers of Abraham and Jacob. The scientific view is that aging, sickness, and death are inadvertent by-products of operating living machines beyond their warranty period.

Nature's lack of intent, and a robust body forged by a harsh environment, create biological loopholes in the contract of life that enable almost everyone to achieve healthier and longer lives. In the absence of genetic programs designed to end life at a predetermined age, there are many ways to favorably influence the aging process and its consequences. The human body is a wondrous amalgam of adaptations characterized by redundancy, over-engineering, regeneration, repair, and the capacity to learn. The following sections illustrate how these biological loopholes can be exploited in order to enhance the quality and possibly even extend the length of our lives.

The IMMUNE SYSTEM

The opening sentence of the first definitive text on immunology—written by Jules Bordet, a Belgian scientist, in 1898—reads: "Life is the maintenance of an equilibrium that is perpetually threatened." When foreign substances or cells invade the body, the immune system is the primary defense against the health threats they pose. As such, the immune system must accomplish the daunting tasks of recognizing "self" from "not-self," alerting the rest of the body that an invasion is taking place, and mobilizing a counterattack to destroy the invaders. Once the invader is defeated, the immune system remembers it, and the body becomes resistant to future attacks by that invader. This remarkable learning mechanism is the basis for the development of immunizations for such infectious diseases as diphtheria, influenza, measles, mumps, polio, and rubella. Childhood immunizations are intended to establish a defense as soon as possible against the many infectious diseases that have historically been the primary threat to health and life at this vulnerable stage of the life-span. Unfortunately, most children throughout the world are not immunized. For this reason, infectious diseases remain a significant and unnecessary source of childhood mortality. In addition, children who suffer from malnutrition also experience a related weakening of the immune system. Survival through the perilous times of infancy and early childhood is a public health issue that can be solved by the universal immunization of children and by attacking the poverty that leads to malnutrition.

The human immune system is incredibly complex, and we still have much to learn about how it operates. One thing that is known, however, is that the immune system is compromised by the same accumulation of genetic mutations that degrade the function of other genetically driven processes in the body. As a consequence, a bacterial or viral infection that is a nuisance at age forty may be lethal at age ninety. Learning how to prolong and possibly even enhance the competence of the immune system would provide an incredibly effective weapon against the infectious diseases that are important causes of death at older ages. Although many technical obstacles exist, the immune system may eventually be enhanced to the point that it can be used to combat cancer and other age-related diseases. For example, it may soon be possible to implant a gene into cancer cells that enhances the ability of the person's own immune system to attack the cancer.

The antibiotics that were discovered in the middle of the twentieth century have saved millions of lives. These miracle drugs have proven to be an extremely effective weapon in the war against the bacteria that historically have often defeated the immune system of humans. Although the war with bacteria will never end, and the recent battles we have won may be only temporary victories, antibiotics have been a welcome reinforcement for our beleaguered immune system. However, the emergence of antibiotic-resistant strains of bacteria has led to concerns over whether the use of these artificial extensions of the immune system can be sustained over the long term. Will antibiotics give rise to more virulent strains of bacteria that totally overwhelm the immune system? Could a continued and growing dependence on antibiotics weaken the immune system? Questions such as these are causing researchers to look for alternatives. Ongoing studies are investigating how nutrition, exercise, stress, previous exposure to viruses and other illnesses, as well as lifestyle factors, can either enhance or detract from the normal functioning and competency of the immune system. Working to enhance the body's own protective mechanisms may prove to be a highly effective method of combating disease, maintaining health, and extending longevity.

MUSCLES

Individuals are born with a finite number of muscle fiber cells. During the course of development, muscle fibers stretch with the lengthening of bones and increase and decrease in size in response to use. Once a muscle cell is destroyed for any reason, it can never be regenerated. As individuals grow older, their muscles begin to lose mass—a process known as atrophy. The good news is that muscles can increase in size and improve responsiveness with use, making it possible to better perform the many activities required for daily living. More importantly, improving muscle fitness enables people to continue exercising at older ages, potentially extending their period of independence. The beauty of exercise is that its benefits are multiplicative—not only do muscles grow larger and more responsive, but the cardiovascular, skeletal, and respiratory systems also benefit. Additionally, exercise can provide benefits at any age; it involves little cost other than effort; and it can be as simple as walking, swimming, or gardening.

Exercise improves health; it may contribute to making people less vulnerable to aging-related diseases and disorders, and it definitely reduces some life-threatening risks among the elderly. For example, the bones of older men and women are often brittle and therefore easily broken. As a consequence, a fall can be a life-threatening event for an elderly person. Those who are more physically fit are less likely to fall.

The improved fitness derived from exercise can dramatically improve the overall quality of life. Through exercise, it is possible to train our bodies at any age to develop a level of physical fitness, as measured by muscle mass and flexibility, that can be even better than that experienced at an earlier age—particularly for those who did not exercise when younger. Athletes are a perfect reminder of the incredible range of potential fitness that exists within every individual. The strength of muscles and flexibility of tendons, within limits imposed by the passage of time, are important aspects of aging that nature permits us to modify. The design and incredible responsiveness of the human body means that at any age, we can all choose to have a stronger, fitter body and an improved quality of life.

BONE MASS

Contrary to their appearance, the bones of humans and other animals with skeletons are in a constant state of flux, continuously breaking down, recycling, and rebuilding. Calcium levels in the bones are regulated by two enzymes, one that releases calcium from bones to the bloodstream and one that performs the opposite function. At two months, the 222 bones of the fetal skeleton are well developed but contain only about 12 percent calcium—the calcium-releasing enzyme keeps the bones in a cartilage phase. Around six months, the calcium requirements of the pregnant mother increase dramatically as the mineralization of the fetal skeleton accelerates. If the mother does not ingest enough calcium, then the fetus will "steal" it from the mother—one of the many reasons why prenatal care is so important for both mother and child. During the first three decades of life, the enzyme responsible for calcium deposition dominates and skeletal mass increases, eventually making up 90 percent of adult bones.

For reasons that are not yet known, the calcium-releasing enzyme gains dominance around age thirty, and a loss of bone mass begins. Over time, the loss of calcium causes the bones to become progressively more brittle, which can lead to osteoporosis, especially among women. Although a loss of bone mass is inevitable, most people die from other unrelated illnesses before it can become a serious health problem. However, when combined with muscle atrophy, brittle bones create the opportunity for one of the most dangerous health crises that confront the elderly: a broken hip. In fact, broken hips have become so common among the elderly that bone density measurements are routinely taken by geriatricians to help monitor, predict, and reduce the risk of bone-related disorders.

The progressive and inevitable degradation of the skeletal infrastructure is, in itself, a sufficient reason why people cannot live for centuries, as the extremists of the prolongevity movement would claim. However, some drugs have been developed to reverse bone loss in adults, and scientists have discovered drug-free methods of slowing down bone loss and lessening the severity of the medical problems it causes. A simple but effective approach is to use diet and exercise to build up as much bone as possible before the inevitable loss begins. The age when the problems associated with bone loss appears is delayed when there is more bone mass to lose. Some food manufacturers have caught on to this trend and are now producing products fortified with calcium. Encouraging the young to adopt a commonsense approach to diet and exercise can dramatically reduce their health risks later in life for diseases and disorders associated with the loss of bone mass.

Diet and exercise can reduce the rate of bone loss even after it begins in adulthood. Weight-bearing exercises such as walking, running, and weightlifting have been shown to reduce the rate of bone loss in adults. Even more important, exercising as an adult reduces the risk of osteoporosis and hip fracture at older ages. These interventions may not reverse bone loss or lead to dramatic increases in life expectancy, but they can improve quality of life. The timing of the buildup and breakdown of the skeletal structure may be under genetic control, but the rate at which these genetic processes occur is another critical attribute of aging that nature permits us to modify at will. Lifestyle choices that are largely within your control *can* make it possible either to avoid or greatly diminish the health risks associated with the inevitable age-related loss of bone mass.

The CARDIOVASCULAR SYSTEM

The cardiovascular system is responsible for bringing oxygen and vital nutrients to cells and taking carbon dioxide and other waste products away from cells. These pickup and delivery tasks are performed by blood traveling through thousands of miles of arteries, veins, and capillaries. To appreciate the enormity of this vascular network, consider that there are about 5 trillion cells in the human body and not one of them is more than a cell or two away from a capillary. Everything travels through the vascular network—the red cells delivering oxygen, the white cells of the immune system, the platelets that form clots when blood vessels are cut, and the chemical messengers that allow organs to communicate with each other. Making sure that everything gets to its destination is the heart, pumping 5 ounces of blood per beat at 100,000 beats per day.

Given this tremendous workload, it is no wonder that diseases of the cardiovascular system are the leading cause of death and disability today. The human body design was established 130,000 years ago, when life expectancy may have been as low as 20 years. Today, life expectancy already has exceeded 80 years in some countries. If people are not surviving beyond their warranty period, then they are certainly close to it. Living longer opens the door to new health crises that develop gradually over long periods of time. Atherosclerosis, a hardening and narrowing of the coronary arteries, is one of these health crises. Symptoms of this disease are rarely seen before age fifty.

Coronary artery disease is not new—signs of the disease have been found in the mummies of ancient Egypt—but it is a disease of long life and therefore more prevalent now than at any earlier time in human history. The putative causes of atherosclerosis provide clues on how to either avoid or reduce

the risk of this disease. The ability to convert food to fat efficiently for long-term energy storage is a genetic legacy from
our ancestors that is useful when food is scarce, but it can be a
liability today. In countries where atherosclerosis is common,
food is plentiful, and so are overweight people. In addition,
many of the foods consumed today are high in cholesterol.
This fatty compound is a necessary component of cell membranes and some hormones, but it can also contribute to the
formation of plaque on the inner lining of arteries, which
obstructs the flow of blood. Laserscopes, angioplasty, and
bypass surgery can repair or bypass arteries that have become
clogged, but it would be preferable to avoid the problem in the
first place. Although drugs can be used to control cholesterol
levels, they can also be modified through diet and exercise.
Exercise not only reduces cholesterol, it also reduces body fat,
lowers blood pressure, and stimulates the growth of capillaries—all benefits that doctors believe will lessen the risk of
coronary artery disease and the angina, heart attacks, and
strokes that are caused by this disease. As with the other systems of the body, a stronger and more fit cardiovascular system is a matter of choice.

DIET

Animals that eat fruit have color vision, carnivores have
canines for tearing meat, and herbivores have molars for
grinding up plant matter. The fact that we humans possess all
of these characteristics indicates that we are omnivores, capable of eating and surviving on just about any kind of food.
Humans also have a robust digestive system that can overcome almost any gastronomical challenge, no matter how
unappealing or unhealthy it may appear to people from

another place or culture. Our generalized eating habits are one of the reasons we were able to invade and thrive in nearly every terrestrial environment on earth—a feat accomplished by no other multicellular species. An omnivorous diet also gives us a variety of ways to obtain the nutrition needed for health and longevity.

Organizations like the American Dietetic Association (ADA) make dietary recommendations that are intended to maximize health and minimize nutrition-related diseases. Everyone requires some minimal amount of protein, carbohydrates, fats, vitamins, and minerals, but an optimum quantity of these nutrients cannot be specified for individuals because every person is genetically unique. Nutritional requirements also change within individuals over the course of their lifespan. Organizations like the ADA are aware of these biological realities, which is why they have developed general guidelines that optimize nutrition for a diverse population.

It is not yet possible to provide a complete genetic profile for individuals, and the linkages between genes, nutrition, and nutrition-related diseases are poorly understood. However, as biomedical researchers learn how to apply the knowledge gained from the Human Genome Project, it may become possible to provide individuals with specific dietary recommendations tailored to their own genetic strengths and weaknesses. Until that time, scientists are searching for generic nutritional guidelines that can benefit everyone. The best progress has been made in identifying aging accelerators—foods that should be avoided in order to decrease the risk of disease and premature death. Known accelerators include excessive quantities of animal fat and processed sugars, food products linked to a long list of lethal and debilitating diseases.

Efforts to identify dietary practices that decelerate the aging process have been less successful. A wide variety of foods has been found to contain nutrients that may reduce the risk of heart disease and cancer—carrots for lung cancer, soy foods for breast cancer, cooked tomatoes for prostate cancer, calcium and fiber for colon/rectal cancer, to name a few. Although there is evidence to suggest that these foods may reduce the risk of some diseases for some people, the genetic uniqueness of individuals makes it improbable that this benefit can be shared by everyone. At this time, the link between specific foods and specific diseases is still too speculative to be useful to individuals. In addition, despite claims made by some advocates of prolongevity, there is no scientific evidence to support the argument that specific foods can be used to make you live significantly longer. Nonetheless, a sensible diet is a personal choice that can produce some generalized health dividends for almost anyone who makes the attempt.

RELAXING LIFESTYLES

According to the actuarial tables that are used to determine insurance premiums, the risk of death doubles about every seven years after puberty. The risks of frailty and disability follow a similar age trajectory. For these reasons, physicians and gerontologists suggest that with advancing age, individuals need to be increasingly vigilant about monitoring their personal health. Beyond the ages of forty to fifty, they recommend more frequent visits to the doctor for disease prevention, and adherence to a more regimented lifestyle of diet and exercise in order to satisfy the special needs of an aging body. Great advances have been made in the sophistication and sensitivity of diagnostic tools that can detect potential health crises at early stages of their development. The earlier a disease process is discovered, the more likely it is that the medical miraclemakers will be successful in their attempts to manufacture survival time and maintain health.

An unfortunate, and unintended, consequence of heightened self-awareness about age-related health issues is that a person can become obsessed, even compulsive, about their personal health. It is this obsession or fear that the advocates of extreme prolongevity prey upon. As the baby boomers approach retirement age, there will be an unprecedented number of older people, most of whom will be acutely sensitive about issues of personal health. This means that in the future the prolongevity industry will flourish even more than it does today. We support those who advocate practical approaches to health and longevity that are based on legitimate science: traditional medicine, alternative medicine, evolutionary medicine, nutrition science, and others. However, we should all be wary of those who distort legitimate science and use pseudoscience or no science in order to create and market false promises and exaggerated claims about health and longevity.

Prolongevity advocates have claimed that with the appropriate lifestyle and health products, aging can be reversed, diseases can be prevented, and we can all live healthy and productive lives for at least 120 years. An unstated message within this promise is that growing old and infirm is a personal failure that could have been avoided had we properly followed the advice of the prolongevity industry. This is a false and misleading trap. The biological reality is that regardless of lifestyle, use of dietary supplements, or any products promoted by advocates of prolongevity, some people will die young, a few will attain extreme old age, and most will die at ages between these two extremes. Within these limitations, everyone can take personal responsibility for his or her health. A sensible diet and exercise program can provide health benefits that are inexpensive and available to almost everyone, a far better option than the false promises offered by the multi-billion-dollar prolongevity industry. The logic that gives rise to this sensible advice also leads to a counterintuitive conclusion: *It is all right to adopt an increasingly relaxed lifestyle as you reach progressively older ages.*

Every person has a unique genetic potential for health and longevity. If you follow a population of humans from birth, you will inevitably observe that the population will dwindle in size as some people die at every age. Some will die from genetic diseases that are currently known and others will die when their longevity potential is reached from genetic causes of death that science has either not yet confirmed or not yet identified. There is scientific debate over the contribution that genetic diseases make to total mortality, but there is no debate that they occur and that more genetic causes of death will be identified in the future. The point is that the people who live to the oldest ages were probably born with a genetic potential for extreme longevity.

Life is a selection experiment for people just as it is for fruit flies in a laboratory or animals in nature. Studies of long-lived populations indicate that the survival gains made by adopting healthy lifestyles diminish with advancing age. Accordingly, we believe that the penalty for relaxing lifestyles also diminishes with advancing age. In *Successful Aging*, Rowe and Kahn note that total cholesterol level is not a significant risk factor for coronary heart disease at older ages. Presumably, those who were susceptible to the damaging effects of high cholesterol have already died. Time reveals the people for whom cholesterol levels were not a relevant risk factor for heart disease. If heart disease does not appear by the time you reach seventy years of age, it may not be necessary to either monitor cholesterol levels or adopt dietary constraints designed to reduce them.

The message from the prolongevity movement is that everyone should adopt the lifestyles they advocate—strict dietary constraints, caloric restriction, hormone injections, vitamin supplements, antioxidants, or other so-called antiaging products. If, as it appears, the health penalties associated with relaxing lifestyles diminish with advancing age, then there comes a point in life when people, with moderation, can afford to indulge themselves without fear of damaging their health or length of life. It may be all right to include foods in the diet that were avoided at younger ages because of concerns about heart disease, stroke, and cancer. We are not advocating a wanton disregard of health consciousness—we are strong advocates for the importance and value of exercise and moderation in diet at any age—but extreme vigilance is not necessary in order to enhance health and quality of life.

The effects of aging often prevent the elderly from enjoying many of the activities they engage in when young. For many people who survive to older ages, preparing or eating food is an important form of pleasure. We suggest that those who have reached advanced ages should enjoy the rewards of having lived a long life. Eat an occasional hot dog, ice cream sundae, slice of cheesecake, dip and chips; put a little salt on those french fries, go ahead and order that cheese omelet and a glass of wine. Just be sure to continue exercising.

OUR PERSONAL RECIPE *for* HEALTH *and* LONGEVITY

The most frequent question our friends and casual acquaintances ask is whether we have discovered the Fountain of Youth. Almost everyone wants to know the "secret" to delaying or reversing the aging process, and some of our friends think that we are secretly working on the problem. Sometimes we just point to ourselves and say: "Would we look like this if we knew the secret to staying young?" Other times we respond to their question with what seems like a humorous answer, but is in fact one that we take quite seriously. There is a Fountain of Youth of sorts that is available to anyone who can afford it, and we recommend it to anyone who asks us for the secret to health and longevity. Our personal recipe includes the following: daily vigorous exercise (30–60 minutes per day); plenty of fruits, vegetables, fiber, and moderate amounts of low-fat protein; a restful sleep every night; an intellectually rewarding, nonstressful job, or no job at all; daily body massage; sex at least once a day; and a regular indulgence in your favorite vice: chocolate, barbecue ribs, you name it. The frequency of the indulgence can rise with advancing age at a rate of one or two per week for every decade lived—that is, one or two indulgences per day by age seventy.

Obviously, not many people have the time or money to follow our recipe—including us—but we have little doubt that it would work. The catch, of course, is that our recipe may not actually make anyone live much longer; but it will certainly go a long way toward enhancing their enjoyment and quality of life.

There are no gimmicks or sugar-coated lies in this book. Flossing teeth will not add years to life for most people, injecting hormones will not stop or reverse the aging process, ingesting vitamins, minerals, and antioxidants will not eliminate aging and disease, you cannot think your way to a 100-year lifespan, a caloric restricted diet will not make anyone live to 120 years who does not already have the potential to live that long anyway, meditating and eating fresh fruits and vegetables will not achieve an ageless body and timeless mind, and taking any of the other remedies currently being peddled by the advocates of extreme prolongevity cannot stop, reverse, or eliminate aging. Short of medical interventions that manufacture survival time, there is very little you can do as an individual to extend the latent potential for longevity that was present at your conception. However, the existence of loopholes within the biological contract of life does mean that there are many things you can do as an individual to reduce your risk of disease, enhance your health and level of fitness, and improve the odds of achieving your longevity potential. You can avoid aging accelerators, exercise throughout life, and adopt dietary habits based on moderation and general nutritional guidelines. By accepting and pursuing a responsibility for personal health, you will receive another benefit whose value is beyond measure—the quality of your life will be dramatically enhanced.

IN THE HOLLYWOOD FILM *Dead Poets Society* (1989), the new English teacher, played by Robin Williams, gathers his poetry class in the hallway of a prep school and leads them to a glass case containing old photographs of students who attended the school long ago. One of the students reads the Robert Herrick quote that begins this chapter:

> *Gather ye rosebuds while ye may,*
> *Old Time is still a-flying:*
> *And this same flower that smiles today*
> *Tomorrow will be dying.*

Like images in a mirror, the young students staring at the old photographs are surprised to see faces that resemble their own. A moment of silence follows as the students gaze at long-gone teenagers just like them—young, full of hormones, and with no sense whatsoever of their own mortality. Tension in the air, Williams with his usual dramatic flair says: "*Carpe diem* (live for the day), gather ye rosebuds while ye may . . . seize the day, boys, make your lives extraordinary." The teacher was encouraging his young students to recognize their own mortality and, like Lucretius, live life not as a losing battle against death, but as a never-ending search for new ways to make the most of life. Morrie Schwartz perhaps said it best in Mitch Albom's book *Tuesdays with Morrie*: "If you're always battling against getting older, you're always going to be unhappy, because it will happen anyhow."

Today, we are faced with tabloid science containing false hope and misleading promises doled out by the modern prophets of prolongevity—most of whom are suffering from aging denial. The advocates of extreme prolongevity are selling slick versions of three-thousand-year-old potions and elixirs to people who, like billions before them, yearn to believe it is all true. At the same time, the genuine science of aging is leading us down a path that is more promising and exciting than any of us can possibly imagine. Those of us alive today are the subjects in this great experiment in biology called aging. Within limits imposed on us by nature, how healthy we are as we age—and perhaps more important, how happy we are during each of our days—is a matter of choice and how determined we are to pursue that choice. With that in mind, our parting advice to you is *Carpe diem*—seize the day.

The last enemy that shall be destroyed is death.

———⟩⟨———

1 Corinthians 15:26

Appendix: The Life Table

	MALES			FEMALES		
Current Age in Years	Years of Life Remaining	Days of Life Remaining	Probability of Living to Your Next Birthday	Years of Life Remaining	Days of Life Remaining	Probability of Living to Your Next Birthday
0	73.5	26,846	99.3	79.6	29,074	99.4
1	73.0	26,663	99.9	79.1	28,891	99.9
2	72.0	26,298	99.9	78.1	28,526	99.9
3	71.1	25,969	99.9	77.2	28,197	99.9
4	70.1	25,604	99.9	76.2	27,832	99.9
5	69.1	25,239	99.9	75.2	27,467	99.9
6	68.1	24,873	99.9	74.2	27,102	99.9
7	67.1	24,508	99.9	73.2	26,736	99.9
8	66.2	24,180	99.9	72.2	26,371	99.9
9	65.2	23,814	99.9	71.2	26,006	99.9
10	64.2	23,449	99.9	70.3	25,677	99.9
11	63.2	23,084	99.9	69.3	25,312	99.9
12	62.2	22,719	99.9	68.3	24,947	99.9
13	61.2	22,353	99.9	67.3	24,581	99.9
14	60.2	21,988	99.9	66.3	24,216	99.9
15	59.3	21,659	99.9	65.3	23,851	99.9
16	58.3	21,294	99.9	64.3	23,486	99.9
17	57.4	20,965	99.9	63.4	23,157	99.9
18	56.4	20,600	99.9	62.4	22,792	99.9
19	55.5	20,271	99.9	61.4	22,426	99.9
20	54.6	19,943	99.9	60.4	22,061	99.9
21	53.7	19,614	99.9	59.5	21,732	99.9
22	52.7	19,249	99.9	58.5	21,367	99.9
23	51.8	18,920	99.9	57.5	21,002	99.9
24	50.9	18,591	99.9	56.6	20,673	99.9
25	50.0	18,263	99.9	55.6	20,308	99.9
26	49.0	17,897	99.9	54.6	19,943	99.9
27	48.1	17,569	99.9	53.6	19,577	99.9

	MALES			FEMALES		
Current Age in Years	Years of Life Remaining	Days of Life Remaining	Probability of Living to Your Next Birthday	Years of Life Remaining	Days of Life Remaining	Probability of Living to Your Next Birthday
28	47.2	17,240	99.9	52.7	19,249	99.9
29	46.3	16,911	99.9	51.7	18,883	99.9
30	45.4	16,582	99.8	50.7	18,518	99.9
31	44.4	16,217	99.8	49.8	18,189	99.9
32	43.5	15,888	99.8	48.8	17,824	99.9
33	42.6	15,560	99.8	47.9	17,495	99.9
34	41.7	15,231	99.8	46.9	17,130	99.9
35	40.8	14,902	99.8	45.9	16,765	99.9
36	39.9	14,573	99.8	45.0	16,436	99.9
37	39.0	14,245	99.8	44.0	16,071	99.9
38	38.1	13,916	99.8	43.1	15,742	99.9
39	37.2	13,587	99.7	42.1	15,377	99.9
40	36.3	13,259	99.7	41.2	15,048	99.9
41	35.4	12,930	99.7	40.2	14,683	99.9
42	34.5	12,601	99.7	39.3	14,354	99.8
43	33.6	12,272	99.7	38.4	14,026	99.8
44	32.7	11,944	99.7	37.4	13,660	99.8
45	31.9	11,651	99.6	36.5	13,332	99.8
46	31.0	11,323	99.6	35.6	13,003	99.8
47	30.1	10,994	99.6	34.6	12,638	99.8
48	29.2	10,665	99.6	33.7	12,309	99.7
49	28.4	10,373	99.6	32.8	11,980	99.7
50	27.5	10,044	99.5	31.9	11,651	99.7
51	26.6	9,717	99.5	31.0	11,323	99.7
52	25.8	9,423	99.5	30.1	10,994	99.6
53	25.0	9,131	99.4	29.2	10,665	99.6
54	24.1	8,803	99.4	28.3	10,337	99.5
55	23.3	8,510	99.3	27.4	10,008	99.5

Current Age in Years	MALES Years of Life Remaining	Days of Life Remaining	Probability of Living to Your Next Birthday	FEMALES Years of Life Remaining	Days of Life Remaining	Probability of Living to Your Next Birthday
56	22.5	8,218	99.1	26.6	9,716	99.4
57	21.7	7,926	99.0	25.7	9,387	99.4
58	20.9	7,634	98.9	24.9	9,095	99.3
59	20.1	7,342	98.8	24.1	8,803	99.3
60	19.4	7,086	98.7	23.2	8,474	99.2
61	18.6	6,794	98.5	22.4	8,182	99.1
62	17.9	6,538	98.4	21.6	7,889	99.0
63	17.2	6,282	98.2	20.8	7,597	98.9
64	16.5	6,027	98.0	20.0	7,305	98.8
65	15.8	5,771	97.8	19.3	7,049	98.7
66	15.2	5,552	97.6	18.5	6,757	98.5
67	14.5	5,296	97.4	17.8	6,501	98.4
68	13.9	5,077	97.2	17.1	6,246	98.3
69	13.3	4,858	97.0	16.4	5,990	98.1
70	12.7	4,639	96.8	15.7	5,734	98.0
71	12.1	4,420	96.6	15.0	5,479	97.9
72	11.5	4,200	96.3	14.3	5,223	97.7
73	10.9	3,981	96.0	13.6	4,967	97.5
74	10.3	3,762	95.6	13.0	4,748	97.2
75	9.8	3,579	95.2	12.3	4,493	97.0
76	9.3	3,397	94.7	11.7	4,273	96.7
77	8.8	3,214	94.2	11.1	4,054	96.3
78	8.3	3,032	93.7	10.5	3,835	96.0
79	7.8	2,849	93.1	9.9	3,616	95.6
80	7.3	2,666	92.4	9.4	3,433	95.2
81	6.9	2,520	91.7	8.8	3,214	94.7
82	6.5	2,374	90.8	8.3	3,032	94.1
83	6.1	2,228	89.9	7.8	2,849	93.5

Current Age in Years	MALES			FEMALES		
	Years of Life Remaining	Days of Life Remaining	Probability of Living to Your Next Birthday	Years of Life Remaining	Days of Life Remaining	Probability of Living to Your Next Birthday
84	5.7	2,082	89.0	7.3	2,666	92.8
85	5.4	1,972	87.9	6.8	2,484	92.0
86	5.0	1,826	86.8	6.4	2,338	91.1
87	4.7	1,717	85.6	5.9	2,155	90.1
88	4.4	1,607	84.3	5.5	2,009	89.0
89	4.2	1,534	83.0	5.1	1,863	87.8
90	3.9	1,424	81.6	4.8	1,753	86.5
91	3.7	1,351	80.2	4.5	1,644	85.1
92	3.4	1,242	78.7	4.1	1,498	83.5
93	3.2	1,169	77.1	3.9	1,424	81.9
94	3.0	1,096	75.4	3.6	1,315	80.2
95	2.9	1,059	73.7	3.4	1,242	78.4
96	2.7	986	72.1	3.2	1,169	76.6
97	2.6	950	70.4	3.0	1,096	74.9
98	2.4	877	68.8	2.8	1,023	73.2
99	2.3	840	67.3	2.6	950	71.6
100	2.2	804	65.6	2.5	913	69.8
101	2.1	767	63.9	2.3	840	68.0
102	2.0	731	62.1	2.2	804	66.1
103	1.9	694	60.2	2.0	731	64.1
104	1.7	621	58.2	1.9	694	61.9
105	1.6	584	56.1	1.8	657	59.6
106	1.5	548	53.9	1.6	584	57.2
107	1.4	511	51.6	1.5	548	54.7
108	1.2	438	49.2	1.3	475	51.9
109	1.0	365	46.7	1.0	365	49.0
110	0.5	183	45.0	0.5	183	45.0

Index